LUIGI MORELLI graduated with a Masters in Environmental Sciences in 1981. He has a passion for social and cultural change, which brought him to live and work in intentional communities serving people with special needs (Camphill International, L'Arche International), or co-housing communities. In his American historical writings he blends the historical and scientific with the imaginative approach of myths, legends and biographies. His other writings concern cultural renewal from a historical perspective and its present impacts on natural science, psychology, social science and literature. His most recent publication is *Searching for the Spirit of the West* (Clairview, 2023). A full list of his books can be found at: www.millenniumculmination.net

THE LIVING CLIMATE

Computer Modelling or Planetary Harmonics?

*Challenging Prevailing Climate Change
Narratives*

Luigi Morelli

CLAIRVIEW

Clairview Books Ltd.,
Russet, Sandy Lane,
West Hoathly,
W. Sussex RH19 4QQ

www.clairviewbooks.com

Published by Clairview Books 2024

A CIP catalogue record for this book is available from the British Library

ISBN 978 1 912992 68 3

Cover by Morgan Creative
Typeset by Symbiosys Technologies, Visakhapatnam, India
Printed and bound by 4Edge Ltd, Essex

Contents

Introduction

'Everything is reversible in Nature and hence there are also atomic energies that are not life-destroying, but life-creating.'

Viktor Schauberger

Understanding the world from a holistic perspective has placed me in the position of seeing polarities at work everywhere. Take a polemic around a personality, choices entrenched around two contrary solutions, or any hot topic in the news. In most instances spiritual science takes me to a place where I can see that the dilemma finds a resolution when we start seeing what one side and the other advance. Most often both sides have something to offer but, from a spiritual-scientific perspective something is missing even if we add the best from both contributions. Sometimes the way out is obvious, for example when it concerns something I have studied in depth. At other times such a resolution can remain a question mark for a long time.

Steven E. Koonin, an unimpeachable scientist who knows the science of climate change from deep within the system, places in a few crisp sentences the dilemma out of which I have been trying to get for the past decade at least:

> Politicians on the right who deny even the basics that science has settled—that human influences have played a role in warming the globe—are not above exploiting climate science uncertainties, offering them as proof that the climate isn't changing after all.
>
> Politicians on the left find it inconvenient to discuss scientific uncertainties or the magnitude of the challenge in reducing human influences. Instead, they declare the science settled and label anyone who questions that conclusion a 'denier', lumping conscientious scientists advocating for less persuasion and more research in with those openly hostile to science itself.[1]

In the first instance all of science is often denied, but this is often only possible to do so by cherry-picking ... the necessary part of

---

[1] Steven E. Koonin, *Unsettled: What Climate Science Tells Us, What It Doesn't, and Why It Matters*, 188.

science that helps support the denial. In the second case science goes often unquestioned, which means accepting everything that goes down in the name of it, whether supported by solid evidence, or not. Belief takes the place of deep scrutiny; in fact 'believing in science' has become a trite, common slogan, one that paradoxically denies the role of science in moving the human being away from blind faith. Others speak of 'the science' as if science always presented but one unified face to the public.

The focus of this book shares little ground with either of these perspectives. The present approach seeks to broaden the scientific perspective; to move away from an interpretive lens which rests on a modelling of the Earth's behaviour and climate according to primary physical parametres alone. Instead it looks at the Earth as a living organism influenced by forces and processes that are much more complex than dualistic thinking can apprehend, and which cannot be reduced to simplistic equations. Understanding at least some of them is not easy. But in the end the effort pays off. We can then start to build an understanding of the whole anew and see the challenge in its connection with the human being, her thinking and her behaviour, not just external factors.

The present work is articulated along two complementary fronts. On one hand a decidedly Goetheanistic/phenomenological approach to world ecology, on the other the gathering of the best that modern science can contribute to the understanding of climate, an approach that surprisingly places the Earth in relation to the Sun and beyond. The second approach encompasses a very broad perspective. The two approaches also admirably complement each other. The first will look at Earth ecology from the perspective of land masses and their relationship to the atmosphere and Sun. The second will move from the broad expanses of the oceans to the atmosphere, cloud formations and their correlations with solar and planetary cycles. From there we will return to a fuller Goethean perspective.

The View from the Earth

For years I had intended to embark on an in-depth study of the work of Viktor Schauberger, but kept postponing it. When in the early nineties I chanced upon the work of Olav Alexandersson, who popularized Schauberger, I was immediately hooked. I even experimented with one of his egg-shaped in-ground cisterns involving reduction processes of organic substance on the land where I lived. But, getting to really know Schauberger is another matter altogether. It is like studying Goethean science all over again, but the difference here lies in getting used to a whole new language, which derives from how the researcher tried to put into words the richness of his perceptions. I tried to go directly to his work but hit a wall. I then turned to those who had synthesized an understanding of his work and started to see some light. I could then return to Schauberger himself for months of prolonged immersion.

The net result of all of this was twofold. When you read Schauberger you are completely changed. First of all, you start seeing things more fully, and understanding the dimension of all that has been inflicted upon Nature. It is sobering to say the least. But, in a second step, if you fully embrace what the Austrian genius says, you start to see the light. Yes, it may be even worse than we think, but on the other hand, the twentieth-century pioneer offered us ways to better understand Nature, and solution after solution in one field after another. It lays with us humans to decide which way to go. We are not doomed, far from it.

I graduated in botanic science and ecology first and finished a Masters in environmental sciences immediately after, at a time in which nobody spoke about climate change. I have enlarged my views about Nature from everything that was offered from Goethean science, from anthroposophical natural sciences and biodynamics. And yet it was only recently, after reading Schauberger and grappling with his writings, that I can say I have had a new understanding not of natural sciences in general, but of the science of ecology in particular.

Climate change theories only entered in the public consciousness in the eighties and nineties. As I was studying environmental sciences between 1979 and 1981 in one of the first European Masters in Environmental Sciences programmes, on the second year it was made available, the topic was neither explored, nor even broached. And yet someone was already perceiving some dimension of it as early as 1931 and writing: 'It should be noted that formidable climatic changes will occur if, as a result of incorrect systems of forest management and river regulation, the orderly formation of clouds is disturbed. Where these systems have been implemented, the number of thunderstorms has consistently decreased, while those that do occur are becoming more dangerous.'[2] In effect Viktor Schauberger was pointing to human interventions that have now impacted and completely modified the cycle of water at a planetary level, and therefore probably affected the proportions of water and CO_2 in the atmosphere. It is no wonder that he already noticed the glaciers starting to retreat in Europe, an effect which he attributed to modern forestry management.[3] Anticipating a central theme, he was pointing to the global modification of the hydrological cycle, a theme overlapping with but also slightly different from climate change as we know it at present.

So how could someone already see with clarity a problem that, save for other people in the fringes, was completely ignored? How could someone armed with pure and direct observation at the local level anticipate global problems that are presently explored through the indirect means of immense stores of recorded data?

The answer, as in many similar situations, is a unique perception of natural phenomena allied to a completely holistic thinking. As we will see, Schauberger, in common with what Rudolf Steiner couldn't tire of repeating, questioned and pointed the finger to the modern way of perceiving and thinking about Nature. In reference to the crisis he anticipated by half a century he wrote in 1933: 'If humanity does not soon come to its senses, and realize that it has been misled and misinformed by its intellectual leaders, the prevailing laws of nature (with poetic justice) will reliably act to bring about a fitting end to this ineptly contrived culture. Unfortunately, the most frightful catastrophes or scandalous disclosures will have to happen before people become aware that it is their own mistakes that have led to their undoing.'[4] But he also cautioned further in the same document: 'Opposition alone, however, achieves nothing. Our youth will achieve any practical success in their struggle only when the *causes are identified and the errors are revealed* that previous generations and we have made, so plunging the world into disaster' (emphasis added).

---

[2] Viktor Schauberger, Callum Coats, editor, *The Water Wizard: The Extraordinary Properties of Natural Water*, 142.

[3] Viktor Schauberger, Callum Coats, editor, *The Fertile Earth: Nature's Energies in Agriculture, Soil Fertilisation and Forestry*, 2.

[4] Viktor Schauberger, *Our Senseless Toil*, 1933, quoted in Alick Bartholomew, *Hidden Nature*, 259–60.

The View from the Solar System

The view from orthodox science is at present a field highly contested and politicized. What would happen if we look at the largest possible picture and let our thinking be illuminated by as vast a horizon of disciplines as possible? Under this lens climate is illuminated by physics, ecology, oceanography, climatology, meteorology, astrophysics and the historical record (paleontology, paleobiology, dendrology, etc.) to name but a few. In fact modern science can take us in two highly contrasting directions: the science of climate models, dominated by physicists, statisticians and modellers, or a new climate science which is the domain of ecologists, oceanographers, meteorologists, and innovative thinkers who start to see the Earth as a living organism.

This amplified scientific perspective has emerged of late, in fact in parallel with the birth of climate change models, and it is casting a new light on hypotheses that are taken for granted. The factors at play in the formation of climate unearth layer upon layer of anomalies and surprises that defy the simplistic thinking we have been predominantly hearing. Here we see cycles, pulses and periodicities that vary from the few years to millennia, which affect ocean currents and weather patterns. Under this light many things that appear new actually have showed recurrent patterns over the space of millennia. The Arctic warming, which has made the headlines more than a decade ago, was not new: the same trend occurred in the 1940s as part of a recurrent 80–year cycle. During the Medieval Warm Period (from around 950 to 1250 CE) the Vikings were able to grow crops and raise cattle in Greenland. The fossil record in the beaches of northern Canada and Greenland 6000 years ago shows that both places must have been largely ice-free in the summer.[5] But then why don't we find anything of the sort in Antarctica? While the North Pole was melting the southern counterpart appeared totally unaffected. This question and similar ones still await more satisfying answers than we receive at present.

Some General Questions

This book wants to explore two central questions. How could both Rudolf Steiner and Viktor Schauberger foresee the global ecological crisis we are facing so far ahead of their time? How do their

---

[5] Peter Taylor, *A Reassessment of Global Warming Theory*, 39.

realizations stand in relation to what we are told today about climate change? And where does the heart of the matter truly lie? Coupled to this is the fundamental framework of reference. Is the Earth a closed system and can it be therefore understood by physical parametres and models alone? Or do we need a wider ecological understanding of Gaia as a living being in constant state of exchange with the wider universe?

To do this we need to venture into new territory. The matter revolves around a couple of important questions: Is it enough to see external correlations and immediately assume that there is a causal connection, or does Nature work in far more complex ways? Is it wise to exclude from our natural science everything that has to do with qualities, rather than quantities? Rudolf Steiner indicated that when science addresses qualities, such as form, enhancement, polarities, subtle variations of one factor or another, it has already moved into what can be called the esoteric, but an esoteric that is derived from a deeper understanding of reality and penetration of sense perceptions, not a desire to move away from the physical and its constraints. All of the above-mentioned factors are continuously at play in Nature and cannot be excluded without losing a deeper understanding of its workings. Steiner was thus implying that it is only by adding a deeper, more truthful and holistic perception of Nature, that we can start to have real answers to global challenges. The hidden dimensions of reality are not a luxury, or the icing on the cake, that we can add after we have proceeded through conventional science; they are the basic prerequisites for a fuller understanding of Nature. Humanity can only exclude them at its own peril.

The journey of exploration we'll embark upon looks at various interrelated aspects of climate change. From the perspective of continental land masses, we will explore water and waterways, forests and farmland and their management, and energy production, in this order. From the perspective of the solar system we will look at ocean currents, solar rays and solar flux, and even peek into influences from beyond our solar system, surprisingly from modern conventional science, no less. We will then explore the growing evidence of Sun and planetary influences in a way that will take us back to Goethean science.

It is the claim of this work that the spiritual aspect of energy can only be understood if we first have a full understanding of world ecology in which water and forest play a universal key role.

And water in the larger sense cannot be understood without a look at its properties and its cycle, that is so crucial in connection with water vapour and CO_2. This cycle is further modified by the planet's relation to the Sun.

What is conventionally accepted as science encompasses only some simplified, quantitative aspects of the whole, but excludes the subtle qualitative dynamics which alter the global picture like night and day. On the heels of a larger systemic and holistic view of Earth ecology and the question of energy, we can create a fuller understanding of the evolution of climate. On the basis of this understanding this book argues that 'planetary depletion' is a better term than climate change when we realize that not only do we have directly visible and measurable changes but also sizeable, qualitative alterations in the being of Nature and Earth itself.

The nature of the crisis looming over humanity will here be called planetary depletion in consideration of the fact that climate change only considers quantitative aspects of the global picture. When we look at it holistically we can perceive that the alteration has not affected just this or that ecosystem or the balance of gases in the atmosphere, but something more pervasive and subtle. It has changed Nature itself to the core. Rivers, forests, farmland and ecosystems are presently different from what they were intended to be. This pervasive change is seldom spoken about, though it's not irreversible, far from it.

A Synopsis

This work will highlight the radically different perspectives coming from viewing the world as a closed system, which logically becomes the domain of physicists, statisticians and modellers, or from the perspectives of a mostly open system, which can only be understood by oceanographers, ecologists, climatologists, etc.

Part I is devoted to a new understanding of world ecology in which the views of Viktor Schauberger and other Goethean scientists form the central contributions. Through these we can assess the true nature and primary cause of the present ecological crisis.

Chapter 1 looks at the view from the closed-system perspective and tests the closed-system hypothesis. In it we will present the place of climate change models, their origins, premises and assumptions.

The concerns of Chapter 2 are the foundations for a phenome-nological exploration of climate change, or in the language of this work, planetary depletion. In order to understand the following chapters it is of primary importance to come to know Viktor Schau-berger, the depth of his work, his methodology and how ideas and practice work hand in hand in this towering giant's work. Every-thing that Schauberger discovered or posited is supported by the effectiveness of his practical applications and technology.

The crux of the matter is entered into in Chapter 3. A purely quantitative and deterministic worldview completely misses the 'being' of water in its crucial role in world ecology. On one hand we have the miracle of a substance that we can never fully know, on the other prosaic H_2O. Herein lies antipodal worldviews. Treating water as H_2O is an emblem, a root symptom of our ecological cri-sis. Humanity has to learn anew to recognize the subtle influences that render water a living being or treat it as a dead shell. Among various factors we will primarily explore the effects of temperature gradients and correlated kinds of motion.

Chapter 4 will look at the land masses, through forests and farmlands. Forest management has resulted in the estrangement of the tree from its environment. Much can be said in the same direction of agriculture. Part of this estrangement of the human being from Nature and of the farms and forests from their envi-ronment is the result of an atomistic thinking, which human-ity had to traverse by way of evolutionary necessity, but out of which it has to emerge for the sake of its future. The completely materialistic, prevailing views lead to disastrous forest and farm management with consequences that can be measured in the worldwide modification of the water cycle.

What is the place of energy production in the whole is explored in Chapter 5. Do we really fully understand what energy is in relation to growth and upbuilding in plants, in relation to emer-gence of new forms? How has it become acceptable to compare what happens in a plant with the workings of an engine? Here we have antipodal worlds. As in Nature so in technology we find ourselves at a great divide. Our generation of energy is contrasted with its polar counterpart, bioenergy. To technological motion we can substitute planetary motion, to centrifugal explosion, centrip-etal implosion. Not only does this come closer to understanding energy in Nature, it also offers revolutionary and abundant sources of energy for humanity's future. We will therefore re-evaluate the place of energy generation in the overall crisis.

From a completely different angle than the previous chapters, it is possible to shed light over a convergent movement in economics. This will be the object of Chapter 6. The work of Gunter Pauli and the so-called 'Blue Economy' has brought ideas similar to those of Schauberger in the domain of the economy. Industrial/technological methods are now predicating the idea of emulating Nature by using natural processes that take place at ambient temperatures, low energy input and generate little to no waste. They offer ways to address climate change at the inception, as it were, and much more efficiently than anything predicated at present.

Part II will in a sense move from the Earth's ecosystems to Gaia's 'planetary ecology.' Once the overview of Schauberger's work is completed we will expand our gaze from the oceans and atmosphere to the Sun, and from the last 150 years to the history of climate over millennia in Chapter 7. We will explore the intricacy and wisdom of an untold variety of cycles. We will once more challenge the view of the Earth as a closed system. This time around it's not Goethean, but conventional science that will show us Gaia in an open relationship to the universe.

On the basis of all that has preceded we will review the science of climate models, the limitations inbuilt in our view of the Earth as a closed system and the consequences of it relating to climate. This will be the object of Chapter 8.

It was the most delightful part of the discovery leading to this book, that we can close the circle, so to speak, in Chapter 9. Not only can we come to question the closed-system perspective from within the Earth's ecology, we can also do so when we look at the rhythms that play out between the Sun and the whole of the solar system, the atmosphere and the oceans. As Schauberger did in looking at the Earth as an ecosystem, so now can we follow other pioneers who look at the Earth as a system attuned to the whole periphery of the solar system and beyond. Two views of the solar system will emerge once more in parallel with the contrasting views of Gaia as a closed or open system. The mechanistic worldview modern civilization has inherited from Galileo and Newton will be contrasted with a return to and further elaboration of Kepler's harmonic understanding of the universe. We will round off our exploration by exploring whether we can predict climate and how it will evolve. Is there a Goethean science to base these predictions upon?

In concluding we will tentatively try to detect similarities and patterns in the broad perspectives outlined in Parts I and II of the book. We will base this on a new understanding of the Sun and solar system.

PART I

CLIMATE CHANGE OR PLANETARY DEPLETION?

THE VIEW FROM THE EARTH

Chapter 1

Two Worldviews

'And for our purpose theories are worse than useless. The world of ideas which comes to light in man must be brought to bear on his perceptions if he is to achieve real knowledge.'

Rudolf Hauschka

Climate change science brings together a multi-disciplinary scientific approach. At the heart of this are primarily laws of physics concerning transformation of energy. Beyond statistical recognition of global human-generated changes to climate and the biosphere lies a view of the Earth as a closed system, though in exchange with the energies of the Sun. The state of equilibrium is altered in ways that can be understood through laws of the transmission of energy, its reflection, absorption and the effect of delayed releases. This is what is known today as 'radiative forcing'.

Radiative Forcing and Greenhouse Gases

Radiative forcing (or 'climate forcing') looks at the flow of energy in the atmosphere, quantified in watts/square metre. It attributes the causes of climate change to either natural or man-made factors. Zero radiative forcing indicates a planet in 'radiative equilibrium' reaching a 'planetary equilibrium temperature'. Positive climate forcing indicates that the Earth keeps accumulating energy over time; it receives more energy than it radiates back in space, because of greenhouse gases, and this net gain drives global warming.

Radiative forcing depends on the added effects of solar insolation/irradiation (the energy per unit area and unit of time received from the Sun in the form of electromagnetic radiation), absorption from dark surfaces (oceans, land to some degree) surface albedo, (or reflection of solar radiation from light sources like clouds or snow/ice) and the atmospheric concentrations of those that are known as greenhouse gases and aerosols. The first ones have a positive effect on the energy balance sheet, the second ones a negative one.

Measurements of radiative forcing, in watts per square metre are obtained at the tropopause–boundary between the troposphere and the stratosphere, situated between 9 to 17 km (5.5 to 10.5 miles) of height—and at the top of the stratosphere, at about 50 km (31 miles).

Climate is what occurs over the space of decades. It takes at least a decade to determine its patterns and two or more to identify the changes. What has triggered the alarm for climate change have been a number of observations of changes over decades, such as the rise of greenhouse gases and a movement toward global warming especially in the years 1980 to 2000. We will review these findings before moving on to areas of agreement and areas that are still questionable. We will do so based on those that are accepted as the most competent worldwide sources, chiefly Intergovernmental Panel on Climate Change (IPCC), National Academy of Sciences, World Meteorological Organization, National Oceanic and Atmospheric Administration and others.

It is possible to pinpoint objective changes in the atmosphere attributable to anthropogenic sources. Let's look at these more closely in the Earth's atmosphere. Nitrogen and oxygen account for 99% of the dry atmosphere: heat passes through them easily. Argon is the next largest constituent. The remaining water vapour, carbon dioxide (CO_2), methane (CH_4), nitrous oxide (N_2O), ozone, intercept 83% of the heat emitted by the Earth's surface. Water vapour is the most important of the greenhouse gases and accounts for more than 90% of the intercepted heat in the atmosphere. After that CO_2, which accounts for 7% of the atmosphere's intercepted heat, and methane. Aerosols play a different role since they increase the albedo or reflective capacity of the atmosphere; they have a cooling effect.

Carbon dioxide is the gaseous compound most affected by human activities. Its concentration has gone from 280 ppm in 1750 to 410 ppm in 2019; during that time the fraction of heat intercepted grew from 82.1% to 82.7%. Carbon dioxide grows on average at a rate of 2.3 ppm/year. Its additional amounts come from the burning of fossil fuels. A doubling of CO_2 from its original baseline, or an amount of 560 ppm, would mean an interception of 83.2% (a further 1% increase from 1750) under clear sky conditions.[6]

CO_2 emissions present the problems of cumulative amounts. Carbon's 'reservoirs' are found in the geosphere in soils and living beings. The amount present in the atmosphere equals 580 gigatons,

---

[6] Steven E. Koonin, *Unsettled*, 51–52.

a 25% of what is in the geosphere, but only 2% of what is in the oceans. What is emitted from fossil fuels amounts to 4.5% of what circulates each year. Half of this amount stays on the surface—partly through added vegetative growth—and the rest goes into the atmosphere.[7] It is quite clear from as many as five lines of inquiry that the increase of CO_2 in the atmosphere is due to human activity. These are:

- The timing of the rise in CO_2 concentrations: the mid-nineteenth century.
- The size of the rise, matching the use of fossil fuels.
- The distribution patterns in the more industrialized northern hemisphere; the rise of CO_2 concentrations precedes the southern hemisphere by two years.
- Tracing of carbon isotopes; of all carbon 1.1 % is the ^{13}C form, while the rest is ^{12}C, a lighter isotope, preferred by living organisms (thus also fossil fuels). We can note that the carbon in the atmosphere has become progressively lighter.
- The very small decrease of oxygen in the atmosphere matches what would have been consumed in burning the fossil fuels into CO_2.[8]

The reason the CO_2 rise raises the greatest concern is that, reputedly, it doesn't dissipate or combine as other gases do. It stays in the atmosphere for a long time. Almost 60% of what is in the atmosphere today will still be there in twenty years and between 30% and 55% will remain in a century. This means reductions in CO_2 emissions will only slow the increases but not the total amounts.[9] This has brought scientists to formulate that our greenhouse gases emissions, chiefly through CO_2, are producing a warming influence on the planet.

Of added concern is methane (CH_4), which has also been increasing over the past century: from roughly 1650 ppm in 1984 to close to 1850 ppm in 2016. Methane has in potency what CO_2 has in duration. It is thirty times more potent in warming, but doesn't accumulate like CO_2. Methane sources are very abundant and diverse, and many of them proceed from agricultural uses (fermentations, rice cultivation, manure,...). Oil and gas use account for only 20% of global methane emissions.[10]

---

[7] Steven E. Koonin, *Unsettled*, 64.

[8] Steven E. Koonin, *Unsettled*, 66.

[9] Steven E. Koonin, *Unsettled*, 68.

[10] Steven E. Koonin, *Unsettled*, 70.

Reviewing scientific literature it is possible to recognize areas where there is a high degree of certainty about the data and agreement between scientists. The findings are reported according to levels of probability: 'virtually certain' (99–100% probability), 'very likely' (90–100% probability), 'likely' (66–100% probability), 'about as likely as not' (33–66% probability), 'unlikely' (0–33% probability), 'exceptionally unlikely' (0–1% probability). A parallel scale, the one more often used, refers to a combination of weight of evidence and agreement among scientists. It will refer to correlations between events and causing factors from 'very high confidence' to 'high confidence', 'medium confidence', 'low confidence' and 'very low confidence'.

Global surface temperature has been evolving at a rate of 1.1°C over 120 years or 0.09° C per decade. But the evolution is not linear: it was of 0.2°C/decade between 1980 to 2020, negative between 1940 and 1980 (-0.05 °C/decade), and 0.09°C per decade from 1910 to 1940.[11] Likewise surface (upper 300 metres) ocean temperature seems to have been rising over the centuries and especially over the last thirty years. Growing ocean heat content is said to be one of the surest indications of planet warming in recent decades.

In relation to temperatures it is interesting to have a closer look. It is noteworthy that the average temperature in the US, where records are very abundant, has moved upward since 1950 but in ways different from those we would expect. Record low temperatures have become more rare, whereas record daily high temperatures have remained stationary.[12] We could say that we are not overheating: we are undercooling!

There is evidence for increased precipitations in the northern hemisphere—US, Europe, temperate Asia—not at a global level however (looking at latitudes from 60°N to 60°S).[13] Through the record of a century of data it is hardly possible to find any significant correlation between climate change and extreme weather events.[14] The IPCC's Fifth Assessment Report (AR5) of 2013 records *low confidence* in relation to trends in the magnitude and/or frequency of floods on a global scale. It restates pretty much the same

---

[11] Steven E. Koonin, *Unsettled*, 31.

[12] Steven E. Koonin, *Unsettled*, 100.

[13] Steven E. Koonin, *Unsettled*, 132.

[14] Steven E. Koonin, *Unsettled*, 98.

about droughts since the middle of the twentieth century.[15] The IPCC's AR5 Working Group I's report attributes low confidence (4 down in a scale of 5) gradings to:

- magnitude and/or frequency of floods on a global scale;
- small-scale severe weather phenomena such as hail and thunderstorms;
- large scale changes in the intensity of extreme extratropical cyclones since 1900 ...[16].

The latest climate assessments from US government and the UN also report:

- Greenland's ice sheet isn't shrinking anymore rapidly today than it was eighty years ago;
- the net economic impact of human-induced climate change will be minimal through at least the end of this century.[17]

Things stand in a somewhat unique situation concerning the rise of sea-level waters. The same AR5 assessment concludes: 'The results are consistent and indicate a significant acceleration that started in the early to mid-nineteenth century, although some have argued it may have started in the late 1700s.'[18] Since the trend preceded the recent acceleration of industrialization it is hardly possible to ascertain how much of the rise in global sea levels is due to human warming or to long-term natural cycles.

Note in passing that the lack of clear correlation in relation to all of the above phenomena does not necessarily mean that they are not happening; it may simply mean that there may be something at play more complex than just CO_2. After all this is the one major correlation fed into and sought by the models.

Climate Models

We will presently turn to the instruments used to assess human influence on climate and predict dangers to come. Essentially, climate models are an extension of weather forecasting. But whereas weather models make predictions over specific areas and short

---

[15] Steven E. Koonin, *Unsettled*, 138.

[16] Steven E. Koonin, *Unsettled*, 98.

[17] Steven E. Koonin, *Unsettled*, 2.

[18] Steven E. Koonin, *Unsettled*, 156.

timespans, climate models are broader in scope and analyse long timespans.

In building climate models and interpreting their results it is important to consider that, no matter their magnitude, human influences constitute only 1% of the energy that flows through the climate system. This means the models have to look at the remaining 99% with a precision higher than 1% (the 'virtually certain' of the scale). This is a great challenge given the available data and the remaining uncertainties about the global system's workings.

Many phenomena, such as the very important differences in heights and coverage of cloud formations, have comparable impact on flows of sunlight and heat as human-generated interferences. Since these take place at very small scales scientists must resort to greater and greater small-scale analysis. Another key uncertainty, which is difficult to integrate in models are contrasting, opposite influences, such as cloud-aerosol interaction. Aerosols have an overall cooling influence contrasting the effect of CO_2.

The logic of the models rests mostly on physics equations, motion of fluids and also some chemistry, important in following the changes in the atmosphere, e.g. O_2 transformed into ozone (O_3). The models work on the fundamental parametres of conservation of mass and energy and how these are exchanged between parts of the system. Apart from the energy balance sheet, climate scientists must incorporate into their considerations temperature fluctuations, wind patterns (speeds and directions), cloud formations and coverage, air pressure, ocean currents, topography, surface characteristics and much more. Increasingly, new factors are added to the pot: the interactions between soil and vegetation (e. g., in relation to forest cover), between land and ice which affect absorption and albedo, and the finer aspects of the carbon cycle between the living beings and the land (biogeochemistry). Models that include this last component are also referred to as Earth system models (ESMs).

The data entered in the models are able to simulate the large-scale evolution of temperature and precipitations, the strength of storms and cyclones, the changes in ocean heat content, the influences of human emissions. Models aim at having both an overall, global accuracy, but also to be predictive tools at a more regional/ local level. The overall complexity impacts predictions and adds levels of uncertainty.

In order to create a model the Earth must be subdivided with a three-dimensional grid of layers of 10 to 20 boxes piled upon each other, their dimensions typically of 100 x 100 km (60 x 60 miles) over land, and 10 x 10 km (6 x 6 miles) over the surface of the oceans. This gives us the head-spinning amounts of 1 million grid boxes for the atmosphere over land, and one hundred times as many over the oceans. The computer programming resorts to the known laws of physics in predicting how air, water and energy move between contiguous boxes over time. Thanks to immense computer capacity the process is repeated millions of times, but even so this may take months no matter how highly evolved the technology.

Over the oceans in order to accurately describe and replicate vertical flows and their variability the boxes have to become more like pancakes than cubes, and even so it is hard to model what happens below the first 10 km (6 miles) where most of the turbulence happens, even more so as we approach the tropics. Consider that what flows from the ocean in terms of energy through evaporation is twenty to thirty times stronger than human influences. Lastly, climate models include a 'time step' of minutes, hours, days, months and years. Here too a trade-off takes place. The shorter the time step the higher the computing power required and the time for processing the data.

Climate models have evolved into three main types: energy balance models, intermediate complexity models, and general circulation models:

- *Energy balance models* basically work on the model of the Earth's energy budget that we have explored. Scientists convert this knowledge into an equation factoring in the energy coming in and the energy going out. A central role in this are the delays in the cycle, such as those due to the absorption of CO_2.
- *Intermediate complexity models* aren't too different from the energy balance models. They present more precision at the topographical level, taking into account geographical features on land, ice features on the oceans. In addition to what the simpler models can do, they allow long-term predictions, and simulate large-scale climate scenarios due to glacial fluctuations, shifts in the ocean currents, or the effects of atmospheric composition changes.
- *General circulation models* are the most complex. They include more specific information concerning the chemistry of the atmosphere, the

variations due to land type, the carbon cycle, ocean circulation patterns and influence of glaciers in any given area. This type of model also uses a grid of higher precision, and requires a larger amount of computing time.[19]

In order to become operational a model must be tested backward, what is known as 'hindcasting'. The goal is to see how well it corresponds to past observed climate data. This serves an additional goal of adjusting the key equations if need be. Furthermore, new models can be compared with existing models worldwide.

Once this look backward is accomplished the model is put to work to simulate future outcomes according to possible scenarios. These are based on choices of population growth, anthropogenic emissions—which condition climate forcing—economic variables, land uses, etc.

Over the years scenarios have aimed at becoming more and more comprehensive. The original SA90 (Scenario A 90) scenarios gave way to the IS92 (IPCC scenario 92) emission scenarios of the 1990s, to the SRES (Special Report on Emissions Scenarios) in 2000 and to RCPs (Representative Concentration Pathways) in 2010.[20] SA90, IS92, and SRES are all emission-based scenarios. They begin with a set of storylines that are based on population projections. By the time of SRES, models had become much more complex, including demographics, trade, flow of information, social, technological, and economic variables.

In much of the above considerations we have been looking at natural ecosystems as extensions of physical systems and models. The large majority of climate change scientists are not biologists, nor ecologists, but physicists. And for good reason since most of the climate modelling is essentially based on physics laws, and chief among them the first and second thermodynamic laws.

In a first instance we can say that they look at the Earth as a closed system, primarily according to the first law of thermodynamics. A simple formulation of the law is: *'the total energy in a*

---

[19] Lauren Harper, *What Are Climate Models and How Accurate Are They?*, 18 May, 2018, at https://news.climate.columbia.edu/2018/05/18/climate-models-accuracy/

[20] Climate Science Special Report, Fourth National Climate Assessment (NCA4) at https://science2017.globalchange.gov/chapter/4/

system remains constant, although it may be converted from one form to another'. Another common phrasing is that 'energy can neither be created nor destroyed', only transformed.

In the conversion of one form to another the second law of thermodynamics predicates that a transformation of one form of energy into another increases the amount of disorder in the system, and that ultimately all energy dissipates into heat. Although energy is indestructible, it changes form when it meets with a resistance, e.g., movement opposed by friction, which generates heat. Thus it is possible and easy to change electricity into kinetic energy and vice versa. These are forms of energy with little entropy/disorder. It is possible to transform electricity into heat, but hardly the reverse. Heat, the most degraded form of energy, which has accumulated entropy, will not generate electricity, other than possibly at very low yields. The second thermodynamic law accounts thus for the irreversibility of natural processes within closed systems.

Various questions arise at this point. Are Nature and planet Earth really only an extension of physical systems and laws alone? Are they really closed systems? Do the facts justify this worldview? To this we turn next.

Matter and Energy: A Permeable Boundary

Conventional science works on a tacit consensus of a widely agreed upon reality. But it rarely satisfies *both* of its two basic tenets. 'Don't accept anything on faith' and 'Don't refuse anything on faith.' The first one is most often satisfied because the sceptic attitude is prevalent in modern time. And there is a great modicum of justification for it. Science after all has replaced blind faith with good reason. It is around the second tenet that the stakes are high, and breaches are common. Fact after fact are roundly denied simply because they don't fit a commonly agreed-upon frame of reference. We will give example after example of these in this book, starting from this chapter. Part of the problem is that science has lost the ability to look at the world in wonder, truly see what is in front of our eyes and then ask the important questions. As a matter of fact many interesting and puzzling facts are known, treated as if they didn't exist and questions are rarely asked. At most they are considered exceptions to the rule ... which don't justify revisiting the rule. But these exceptions are legion. This places science at the risk of becoming another belief system, counter to what its role was in the first place.

In what follows we are going to refute many hard-to-dispel myths of modern science, myths that have been disproved for more than a century or two in various instances, but still form the unquestioned bedrock of our set of scientific assumptions. To do this we will introduce the work of various, known or forgotten scientists and then see how their findings are elucidated by the most recent phenomenological, or spiritual-scientific researchers.

Is Nature, or any part thereof, really a closed system? Many experiments and observations roundly deny this assertion. Matter can be created 'ex nihilo'—a bold assertion which can be proved—thus contradicting the assumption of Nature as a closed system. These things have now been put to the test for centuries, only to be roundly denied … because they fall outside of the purview of the tenets of thermodynamics and of closed systems. They are known to many who study Goethean science and the work of spiritual science.

The canon of science lies in being able to submit what we want to know to experiments and reproduce the results of these tests over time, no matter how puzzling or unsettling they may be. This is what scientist Rudolf Hauschka did early in his scientific career. Having come across the work of Otto Philipp Albrecht von Herzeele (born 1821—date of death unknown) who wrote *The Origin of Inorganic Matter*, he simply set out to lay the foundations of his scientific understanding by checking the truthfulness of his predecessor's claims. He replicated his experiments.

Hauschka must have been a very determined scientist because he came in possession of the apparent last copy of von Herzeele's work. So what did his predecessor demonstrate? In several hundred experiments conducted between 1875 and 1883 von Herzeele placed seeds in distilled water and sealed them in porcelain bowls covered with glass bells. The air was filtered in order to remove all dust. According to the tenets of conservation of matter and energy, no matter could accrue from the plant other than what was present in the seed itself. In spite of this the analyses of the ashes of seeds and plants brought out an increase in all chemical elements.

Fine-tuning his work von Herzeele then replaced the distilled water with particular saline solutions. Here are some of his unexpected results. When a phosphorus solution was provided, contrary to the expectation of higher percentage of phosphorous in the ashes, it was the sulphur content which increased dramatically, leading him to the first realization that plants can transmute phosphorous into sulphur. A series of similar results ensued. Out of a calcium saline solution plants increased their amount of phosphorous. For calcium to increase one can place the seeds in a magnesium salt solution. And the magnesium content was increased in the presence of water with carbonic acid in solution. The scientist thus found the sequence of conversion CO_2 (or HCO_3)→magnesium→calcium→phosphorus→sulphur. On the other hand he also saw the conversion of nitrogen into potassium. It seems Nature is not such a closed system, not even in the most artificially closed-off setting.

With the advantage of even more developed laboratory settings—sealed ampoules which could also prevent the entrance of atmospheric gases—Hauschka carried out experiments between the years 1934 and 1940. In what could not be a more perfectly closed system he was able to confirm that '[von Heerzele's] findings must therefore be extended to the statement that plants not only generate matter out of a non-material sphere, but under certain circumstances again etherealize it'. In certain situations in effect, Hauschka was able to record decreases of weight, thus the disappearance of matter. The greatest majority of the negative results came during the new moon phase.[21]

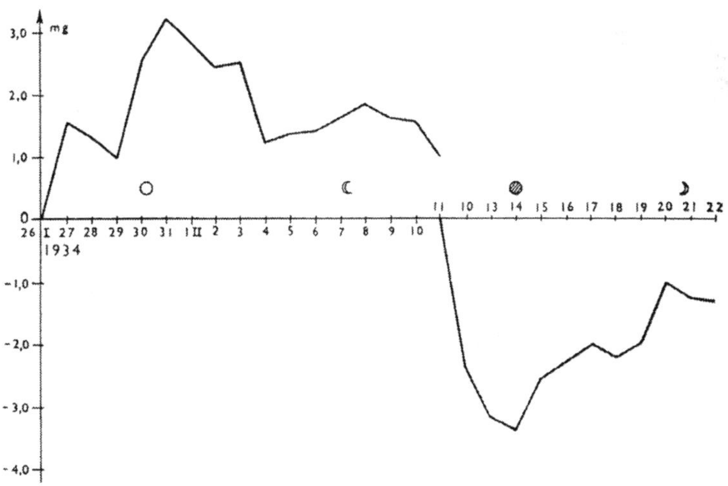

Figure 1: Weight changes of sprouting seeds in a closed system

---

[21] Rudolf Hauschka, *The Nature of Substance: Spirit and Matter*, Chapter 3: 'New Ideas on the Primality of Spirit.'

Schauberger too recognized this larger reality of the conversion between energy and matter under various aspects. 'Everything that appears in Nature and is perceptible to our eyes and senses, is the waste product of subtle exalted energies—left behind, foot-sore and weary on evolution's upward path, it manifests itself as physical matter.' An example of this for the Austrian scientist: 'oxygen is waste-matter or fallout of solar energy in gaseous form'.[22]

Does Gravity Operate Alone?

Our understanding of climate change also rests on the unquestioned laws of gravity, since no force opposed to it is known to science to this day. This is another long-held denial of what stands in plain sight and has been so for a long time, since it was at the end of the eighteenth century that Samuel Hahnemann discovered the principles of homeopathy, and in 1810 he published his seminal Organon of Rational Medicine. His discoveries should have shaken the scientific consensus to the core.

It was Rudolf Hauschka once more who brought back scientific facts already known through homeopathy and tested them in new ways. Homeopathy utilizes repeated 1 to 10 dilutions of plant, mineral or animal substances alternating with rhythmic potentizing of each new solution in order to increase the efficiency of the remedy. The simplest potentizing in Hahnemann's time was the 'succussion', a simple shaking of the solution; at present potentizing uses rhythmic movements. The rhythmically potentized solutions act differently from simple, progressive 1 to 10 solutions with no potentizing in between; the latter ones simply lose more and more of their effect until it completely vanishes.

At homeopathic dilutions of 1: 10^{17} no detectable substance is left in the solution and in the resulting remedy. And yet, the remedy is still active, at times even more so. In his time Hahnemann proved that there was a spiritual principle at work in matter, and that principle is what renders the remedy efficient, while substance vanishes. We know from spiritual science that to each substance corresponds the work and imprint of a formative force. Rudolf Hauschka set out to prove that chemical compounds are different whether the molecule is produced synthetically, or by a

---

[22] Viktor Schauberger, Callum Coats, editor, *Nature as Teacher: How I Discovered New Principles in the Working of Nature*, 137.

living being. He studied the activity of yeast and its production of carbonic acid; we know that yeast produces both alcohol and carbonic acid as end products of fermentation. He then plotted the potency curves of various substances, meaning the variations of H_2CO_3 production in relation to progressively potentized dilutions (increases by steps of a 10 multiplier).

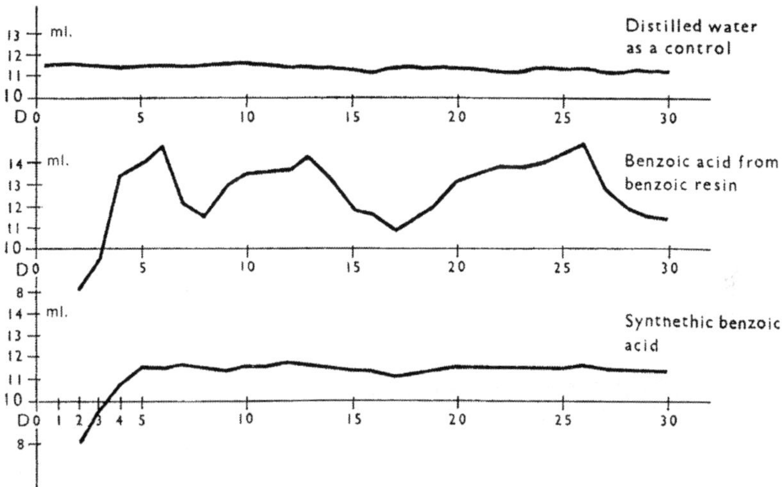

Figure 2: Potency curves of natural and synthetic benzoic acid

In some experiments Hauschka compared the results obtained with two kinds of benzoic acid; the first extracted from benzoic resins and purified; the second synthetically produced. The two are chemically the same, impossible to differentiate. The curve of the natural acid created the well-known result of potency curves with marked minima and maxima. The synthetic acid registered no variations whatsoever; after the fifth dilution; the effect simply vanished. This different behaviour of the two substances explains why synthetic drugs can only work in allopathic dosages. Hauschka concludes 'in the second realm [synthetic substance] the laws of physical atomic and molecular chemistry prevail. In one case we are dealing with organisms, in the other with mechanisms. Hence the law of the conservation of matter is entirely valid in the mechanical realm, but not in the organic.'[23]

---

[23] Rudolf Hauschka, *The Nature of Substance*, Chapter 17: 'High Dilutions and Their Effectiveness.'

To the force we know as gravity corresponds another one that is the carrier of life and that has been called 'levity' in spiritual science. Viktor Schauberger called it, among other things 'levitation'.[24] In his estimate, as in that of spiritual science, both gravity and levity are terms of a polarity; they take place simultaneously. Gravity, however, only holds sway within limited conditions, within the Earth's atmosphere. Beyond it we find mostly or exclusively levity; witness the laboured movements of the astronauts on the Moon in the absence of gravity. This leads Schauberger to conclude: 'Gravitation, however, is only a secondary effect of this principal force [levitation]. Were there no levitation, then gravitation could never exist, because in the final analysis everything would remain flat on the ground.'[25] Newton's apple would not be able to fall.

The contradiction of physical laws can be pushed further in the workings of Nature, at least when in its pristine states. When Nature works at its best, gravity and the Law of Archimedes can be blatantly denied universal validity; witness some of Schauberger's observations in Austrian ecosystems and demonstrations in man-made technology. The Austrian observed in primeval environments that under the full Moon or close to it, some stones, those that were egg-shaped, started gyrating at the bottom of a creek, then slowly came up to the surface. Something similar happened with heavy logs. Cold water and the light of the Moon somehow energized logs and stones, impressing upon them levity of such strength that gravity was overcome.[26] Schauberger concluded in relation to the role of very cold temperatures: 'This cold concentration gives rise to the emission of expansive emanations, which lead to the creation of the "original form of motion". It is this force of movement that overcomes gravity and raises the specifically heavier stones to the surface of the water.'[27]

---

[24] Part of the difficulty of following Schauberger's work lies in the evolution of his terminology. He knew that existing language could hardly be applied to unexplored and unrecognized phenomena, so he adapted and evolved his own terminology and created words. Some of these evolved over time with his own evolving understanding. Levitation is in other places called 'levitism' or 'bio-magnetism'.

[25] Viktor Schauberger, Callum Coats, editor, *Nature as Teacher*, 143.

[26] Viktor Schauberger, Callum Coats, editor, *Nature as Teacher*, 43–45.

[27] Viktor Schauberger, Callum Coats, editor, *Nature as Teacher*, 44.

We will return in the next chapters to what he calls 'original form of motion'.

Schauberger's forefathers already designed log flumes for the transportation of timber in hard-to-reach areas in the Austrian mountains, making the exploitation of timber more financially sustainable. The log flumes were constructed to bring down wood in a trough, whose meandering path mimicked that of a river. They had already noticed that transporting logs, even those heavier than water, was much easier and effective at night, particularly at the times of full Moon. They knew that water was much more alive in these conditions. This is something we will bear out in the rest of our explorations.

Subtle Influences: Form, Materials, Sound and Temperature

We have pointed out in passing to the effect of a particular form, the egg-shape. A science that looks at quality will also be curious about the variable of form, which is hardly considered in conventional science. The question may in fact easily arise as to why some forms are common in Nature, and what may be their function, or advantages. Why can they be used to obtain effects on water or soil quality?

Water is an endless source of wonder, which allows us to capture those finer influences that, no matter how much they be present, can still be ignored and dismissed. The work of Masaru Emoto has popularized how water acts as a storage of memory of all kinds of external impressions, be they chemical, etheric or even soul-related.[28] Rudolf Steiner at the beginning of the twentieth century has recommended to inquire into the etheric realm with a whole new methodology, that of 'rising pictures methods'. In the realm of the etheric we cannot measure as we do in the physical, but we can offer an image of the overall vitality of a medium.

In Steiner's time the capillary dynamolysis method (also called chromatography) was developed and so were crystallization tests using copper chloride. The patterns of diffusion in one, or of the forming of crystals in the other, inform us of the overall strength/vitality of a specific water, or other mediums like human blood or plant sap. More recently the drop picture method, developed by Theodor Schwenk, has been used to detect pollution in rivers or the

---

[28] See for example: Masaru Emoto, *Messages from Water*.

influence of different chemicals on water quality.[29] A clearly formed image will correspond to a healthy water, sap, blood or other liquid. In other words, we can have an overall image of the vibrancy of the liquid medium and its state of health.

The work of Masaru Emoto uses yet another picture test, that of water crystals formed at very low temperatures. The Japanese researcher has popularized how much water is susceptible to all sorts of influences. To name a few: water responds to chemicals, so much so that it may still carry their information even after they have been treated and it is considered legally fit for human consumption. Water reacts differently to various kinds of music, both positively and negatively (absence of crystals), to words spoken or written, to human intentions/activities such as prayer.

Relating to the question of form, straight pipes of glass or copper, and spiral-wound copper pipes (similar in shape to the horn of a kudu antelope) have been tested by Schauberger in relation to water's movement. Friction was progressively lower in the order listed above. In the spiral-wound copper pipe friction was greatly reduced and even reached zero and negative values. In the straight pipes water has the tendency to generate wave movements; the spiral pipe allowed waters to flow freely.[30] The experiments thus brought to light another qualitative aspect that science completely discounts. Why would different materials have different effects: why did copper behave differently from glass?

It is precisely in conjunction with these two matters, form and choice of material, that Schauberger obtained some astounding results in the agricultural field. Form can be determining in more than one way. It may be an important choice even in the matter of farming yields. One of the most successful devises Schauberger developed was a plough of unique form and composition, one that came to be known as the 'golden plough' or 'bio-plough'.

We are talking about a spiral plough plated with copper (see Fig. 7, p. 43). The form of the plough allows for a rounder movement of the soil clod than is possible with conventional ploughs. In field trials it was proved that the difference of form and mate-

---

[29] Jochen Schwuchow, John Wilkes, Iain Trousdell, *Energizing Water: Flowform Technology and the Power of Nature*, 60–61.

[30] Olof Alexandersson, *Living Water: Viktor Schauberger and the Secrets of Natural Energy*, 116–19.

Figure 3: Flowform patterns

rial produced remarkable increases of up to 30% in yield. These also had great effects on pest control through a decrease in parasites.[31] The success of the ploughs is due in no small measure to the substitution of iron with copper, of which more will be said later.

Closer to modern time many who have been interested in biodynamics, or in biotechnology, know of the use of flowforms, a succession of basins of varying shapes used for the treatment of water and added artistic benefit. What they have in common beyond their different shapes is the creation of a lemniscate movement at each basin, and the accrued effect of the succession of basins. Here too a carefully devised form allows the water passing through it to be potentized, therefore enlivened. In effect the flowforms can be used in the last stage of urban wastewater treatment (tertiary treatment), to render water newly drinkable. They can be used in farming for the stirring of biodynamic preparations, or for the health of animals, and for the treatment of pools in such a way that use of chlorine can be completely avoided.[32]

What is at work behind the golden plough, or the flowforms, are of the greatest interest for our exploration. Forces working from a centre to the periphery, also called centrifugal, are those most commonly known; they explain gravity. The forces working from the periphery toward a centre, what we have called levity, can be studied through projective geometry. George Adams rediscovered the 'path-curve' geometry of Felix Klein and Sophus Lee (nineteenth

---

[31] The experimental set-up and its results can be seen in Viktor Schauberger, Callum Coats, editor, *The Fertile Earth*, 185–95.

[32] Jochen Schwuchow, John Wilkes, Iain Trousdell, *Energizing Water: Flowform Technology and the Power of Nature*, 43–44, 93.

century). These are curves that describe the forms of vortexes, pine cones, and the egg-shape among others. Lawrence Edwards' work proves that the forms of buds and cones can be described very precisely with this geometry. In fact the forms of vortexes, eggs and anything egg-like or cone shapes, 'are mathematically closely related and can be transformed into each other by changing only one parameter'.[33]

The path-curves are incorporated in the shape of the flow-forms. They cause rhythmic, lemniscate movements. 'In this way the mountain stream and the living pulse, as nature's own methods for improving water's capacity to support life [quality] are combined in a technology derived from a close study of nature's creative processes.'[34]

Nature has little use for squares and cubes, rectangles and cylinders. In Nature the straight line or the circle are rarely found. The Euclidian system based on points, lines, circles and ellipse cannot comprehend what operates in the upbuilding natural cycles. Nature operates in non-Euclidian geometric systems. Central to these is the golden ratio equation ($a/b = a+b/a = $ phi or φ, the golden ratio), where $\varphi = 1.618...$ an irrational number. This finds an expression in the $\sqrt{3}$ rectangle (derivable from the form known as the 'Vesica Piscis' at the intersection of two circles).

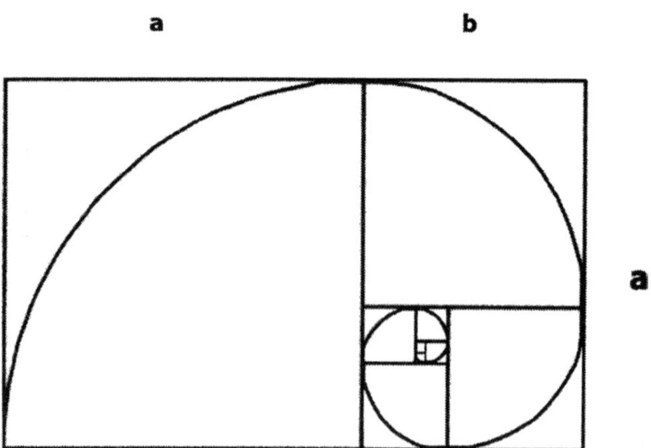

Figure 4: Golden ratio progression generating a spiral form

---

[33] Schwuchow, Wilkes and Trousdell, *Energizing Water* 24.
[34] Schwuchow, Wilkes and Trousdell, *Energizing Water*, 30.

With a series of rectangles enlarging in size according to the proportion of the golden mean, we can construct the spiral forms present in Nature (see Figure 4). Among others this can be seen in the proportions of the nautilus shell, in the head of a sunflower or of a Romanesco broccoli.

Since Darwinian biology has little understanding of the qualitative it sees no function in natural forms, other than haphazard evolutionary results, behind which little meaning can be found. Lawrence Edwards concludes otherwise: 'One finds in Nature a continuous urge towards ever increasing complexity which is completely unexplained by the law of the survival of the fittest.'[35] And many of these forms can arise through the simplest linear processes imaginable. Through the linear process of collineation—transformation of a surface into itself through lines originating from two invariant points, or of a volume into itself through three invariant points—so-called 'path-curves' are obtained through relatively simple geometrical and mathematical processes. Their mathematical equations rest on a key parameter, Λ, which, within a family of path-curves determines the shape of the resulting forms.[36]

One of the most common forms in Nature, the egg, arises from such simple transformations; it is one manifestation of a family of path-curves. Path-curves give rise to families of forms, which are quite remarkable since Edwards aptly reminds us 'their possibilities are strictly limited. The things they cannot do, and the kinds of forms which they cannot assume, are far more numerous than those which they can.'[37]

Within the realm of path-curves two primordial forms which will appear in this work, have a remarkable kinship: the egg and the vortex. In mathematical / geometric terms, a 'negative' form of the egg, obtained by transforming the key parameter Λ, is the vortex. A Λ value of 1 gives us the egg form, a value of $\Lambda = -0.5$ generates the vortex. Above 1 the egg acquires more pointed apices; below 0 we see the vortex gradually arise. Families of path-curves have been found to closely match with significantly high statistical certainty the forms of not just eggs, but also of buds and pine cones.

---

[35] Lawrence Edwards, *The Vortex of Life: Nature's Patterns in Space and Time*, 24.
[36] Lawrence Edwards, *The Vortex of Life*, 54.
[37] Lawrence Edwards, *The Vortex of Life*, 43.

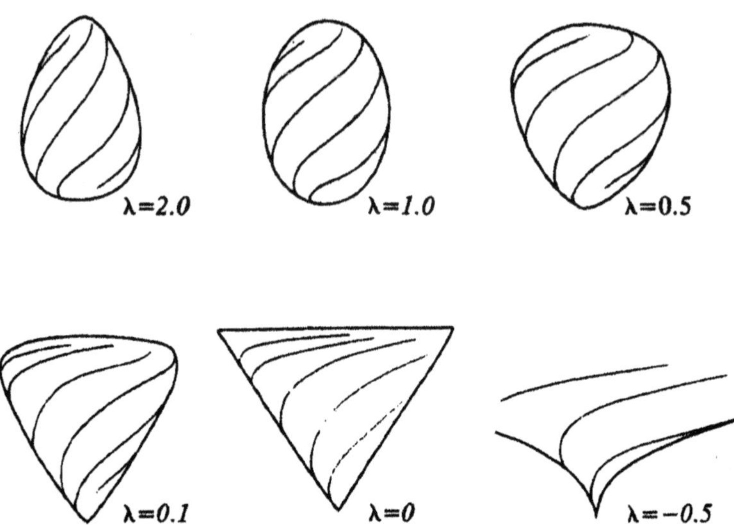

Figure 5: From egg through vortex in relation to the parameter ʎ

The egg-shape is that container which allows water's constant movement especially when confined in materials such as wood, stone or terracotta. This is why Greeks and Romans used amphorae—an extreme egg-shape—sealed with beeswax. Water in these containers has to be stored away from light, in opaque, porous containers. Terracotta allows evaporation through its walls, which also means cooling. Even if just a 600th part of water is evaporated this corresponds to a 1°C cooling. Evaporation gives rise to the slight differences of temperature in the remaining water and therefore a temperature gradient, that generates movement. Through its natural shape the whole body of water in the amphora was kept in movement. This movement occurs also in the medium of air; this is why grains found in amphorae are still viable when found more than two thousand years later.[38]

To end we will close the chapter with one more example of how a minor quantitative variation can affect an important qualitative change. In his youth Schauberger, following a multitude of observations, asked himself: 'How does the trout reach a place above a waterfall of 60 metres (~200 feet)?' Suspecting it had to do with the temperature gradient he set up a simple experiment. He let out 100 litres of hot water some 150 metres upstream from

---

[38] Viktor Schauberger, Callum Coats, editor, *The Water Wizard*, 9–10.

the trout. Even with such a minute change of temperature the trout was unable to maintain its position in the stream and jump upward.[39]

The last example is crucial for our explorations. Water is everywhere present in our environment, from rivers to lakes, seas and oceans, glaciers and icebergs. It is a universal solvent, which defies understanding when compared to all similar substances, one that is incredibly sensitive to all possible, minimal external changes. Part of the climate crisis that we face has to do with the disturbance of its cycle, which will take some time to bring to understanding in the following chapters. The reason for this is that it requires looking at qualitative variables that are completely ignored in conventional science, so much so that our civilization continuously compounds the problems it is trying to solve for lack of a larger understanding. Most of the time it confuses causes for consequences. As Callum Coats, curator of Schauberger's work, reminds us 'quality is the differentiator and animator of life'.[40]

In our present approach to science quantity is paramount, quality optional or unknown. However, for a true understanding of Nature, quality plays a defining role, quantity is subordinated and relatively unimportant. This is what will emerge more and more clearly in the next three chapters.

---

[39] Alick Bartholomew, *Hidden Nature*, 16.

[40] Callum Coats, *Living Energies: Viktor Schauberger's Brilliant Work with Natural Energy Explained*, 64.

Chapter 2

The Water Wizard

'How can it be easy to understand Father Schauberger's language—his work belongs to the future.'

Wilhelm Balters

'We must learn entirely anew. With hands empty and ready to receive, we must climb up into the mountains towards the dawn. … Every marvel, however, is none other than the result of conformities with a law which we still do not understand, and are thus not able to set in motion.'

Viktor Schauberger

Although the first part of this work leans heavily on the phenomenological ecology of Viktor Schauberger, it does not attempt to do justice to the complete work of this towering individual. Rather, it looks just at the essential aspects of the picture in relation to the question of climate change. I refer the reader interested in knowing more to the works of Olof Alexandersson, Callum Coats, Alick Bartholomew and Jane Cobbald, quoted here and there in this work.[41] I'm in fact deeply indebted to their synthesis, which has rendered Schauberger's work more accessible once I went back to the original.

Every now and then a genius is sent to humankind to blaze a trail way ahead of everybody else. Anthroposophists interested in science know of the work of Johannn Wolfgang von Goethe or of Rudolf Steiner. Such individuals appear on the scene of history at important turning points and in places in which change can happen. We can do no other than notice and wonder how both these individuals were born in the culture of Central Europe. So is the following individual to which we turn next, whose deeper understanding of ecology is central to the discoveries of this book. The Austrian was one of the first to offer the science of ecology a wider

---

[41] Olof Alexandersson, *Living Water: Viktor Schauberger and the Secrets of Natural Energy*; Callum Coats, *Living Energies: Viktor Schauberger's Brilliant Work with Natural Energy Explained*; Alick Bartholomew, *Hidden Nature: The Startling Insights of Viktor Schauberger*, and Jane Cobbald, *Viktor Schauberger: A Life of Learning from Nature*.

and more holistic perspective and the first to observe the signs of climate alteration and describe their possible, future evolution.

Viktor Schauberger's Biography and Turning Points

The Austrian pioneer showed us another way of understanding Nature's economy. He rendered obvious how other forces are at work than gravity, the conservation of energy, Archimedes' Law, or other commonly known physical laws. For Schauberger entropy and the law of conservation of energy only apply to some degree in natural systems; they are really meant for mechanical, closed systems.

Schauberger deplored many of the practices that are today prevalent in forestry, agriculture, water management, and all of technology, many of which are hardly questioned even in the ecological movement. In every one of these fields, however, he did not preach a return to the past, other than where it is absolutely indispensable. He delivered solutions after solutions in all realms of Nature's management and technology. He offered us a more complete understanding of water and rivers, forestry and farming, and amazing, if discrete, insights into the whole field of biology, from blood circulation to the physiology and growth of trees, to the particularities and uniqueness of one or another animal.

On the practical end he devised technologies to regulate waterways with the minimal possible intrusion, others to improve the working of dams to lessen their impacts and prolong their resilience, machines for the production of the highest quality water, implements, practices and devises for farming. Lastly he developed a consistent approach to the biotechnology of energy generation.

Fate offered Schauberger the possibility of working in Austria before its primeval forests were more and more abused and degraded. He inserted himself in a long line of people devoted to Nature. He was born on 30 June, 1885 in Ulrichsberg, Upper Austria near Linz, and grew up in an area of original forest, which is something most rare at present in Austria or most of Europe.[42]

---

[42] Schauberger's biographical information has been gathered from Olof Alexandersson, *Living Water: Viktor Schauberger and the Secrets of Natural Energy*; Callum Coats, *Living Energies: Viktor Schauberger's Brilliant Work with Natural Energy Explained*; Alick Bartholomew, *Hidden Nature: The Startling Insights of Viktor Schauberger*; Jane Cobbald, *Viktor Schauberger: A Life of Learning from Nature*, plus Schauberger's own recollections in the four volumes edited by Callum Coats.

The Schaubergers originated from a noble family with a castle in Schauberg, north-east of Salzburg. After falling from grace, a lone remaining Schauberger survivor was banished to the forests around Dreisesselberg, not far from Salzburg to the south. From him came a long line of forest wardens, foresters, gamekeepers, living in pristine lands for almost a thousand years. This meant a long line of people used to live with Nature and observe it keenly, developing intuition and insights. They were true to their coat-of-arms 'Fidus in Silvis Silentibus'—Be faithful to the silent forests.

Although he had many siblings—he was one of nine—with whom he got along well, Viktor preferred to play by himself in Nature, never tired of observing one thing or the other. The child learned a lot from his father and relatives, most of them foresters. The father taught him that water sleeps during the day and wakes at night, especially under the full Moon. His forefathers had already designed log flumes for timber transportation. In some of these it was possible to see wood floating upwards, defying gravity: some of the timber were specifically heavier than water.[43] The youth's ancestors also knew that irrigation water channelled at night would give greater yields than what was done conventionally in neighbouring land during the day.

Viktor felt closer to his mother, rather than his father. The following gives an idea of the old lore and understanding of Nature spirits which Schauberger received from his mother, regardless of whether it can be cosidered correct or not. The four-year-old Viktor had fallen into the waters of a spring. The mother rebuked him thus:

> You silly boy! How dare you go to the water! The poor souls of the departed migrate through the water towards the mountains and are resurrected at the springs and carried up to heaven by the *ur*-force of all life. They entrance you, pull you in, you drown and die, and then you have to go with them. When you are grown up, but not before, you can go to the water's edge. When you're older, and you have pressing problems and no longer know what to do or how to help yourself, go to the waters of mountain springs. There you will find me again, for there my soul will be. I will then give you motherly advice and help you when I'm no longer on this Earth.[44]

---

[43] Viktor Schauberger, Callum Coats, editor, *Nature as Teacher*, 50.

[44] Viktor Schauberger, Callum Coats, editor, *Nature as Teacher*, 30.

Further referring to the gift of his inheritance Schauberger was very aware of his good luck; he realized how close it set him to the kind of knowledge he needed and later acquired. He reflected: 'This person [blessed by such inheritance] need not speculate, because he can see the difference between ancient and modern knowledge and can therefore choose between knowledge and science.'[45] About his life-long search he further wrote: 'Even in the earliest youth my fondest desire was to understand Nature and through such understanding to come closer to truth; a truth I was unable to discover either at school or in church.'[46]

In his childhood Viktor wanted to simply be a forest warden like all his ancestors. The father, however, wanted him to take an arboriculturist training. He only did it briefly, but in the end graduated with the state warden's exam, further resisting his father's entreaties. He acted thus because he had seen what academic teaching did to those around him, especially his two elder brothers who had gone to university. He sensed how much their training had altered their way of thinking. He refused college education because he feared he would lose his intuition and his perception of the magic at work in Nature. He later wrote: 'The only possible outcome of the purely categorizing compartmentality, thrust upon us at school, is the loss of our creativity. People are losing their individuality, their ability to see things as they really are and thereby their connection with Nature.'[47]

Viktor appeared dreamy due to his psychic and intuitive abilities, but he also had a keen intellect and a gift for very precise observation. He awoke to the elemental being of water and could say about this experience: 'I could sit for hours on end and watch the water flowing by without ever becoming tired or bored.' It was not long before Schauberger developed a unique way of understanding and studying Nature, originally in relation to water, his first and long-lasting love. He explains:

> Gradually I began to play a game with water's secret powers; surrendering my free consciousness and allowing the water to take possession of it for a while. Little by little this game turned into a

---

[45] Quoted in Olof Alexandersson, *Living Water*, 125.

[46] Viktor Schauberger, Callum Coats, editor, *Nature as Teacher* 29.

[47] Viktor Schauberger, *Our Senseless Toil*, Part. I, 28–29, quoted in Callum Coats, *Living Energies*, 3.

profoundly earnest venture because I realized that one could detach one's own consciousness from the body and attach it to that of the water. When my own consciousness eventually returned, the water's most deeply concealed psyche often revealed the most extraordinary things to me. As a result of this investigation a researcher was born who could dispatch his consciousness on a voyage of discovery. I was just able to experience things that had escaped other people's notice, because they were unaware that a human being is able to send forth his free consciousness into those places the eyes cannot see. By practicing this blindfold vision, I eventually developed a bond with mysterious Nature, whose essential being I slowly learnt to perceive and understand.[48]

The young Austrian soon married and apprenticed as forest warden not far from his family of origin. He was drafted four weeks after the birth of his first child in 1914, fought in WWI in Russia, Italy, Serbia and France and was wounded. After the war he became first forest warden then gamekeeper and went to work for Prince Adolf zu Schaumburg-Lippe in the pristine 21,000 hectares domain in the forest of Bernerau in Steyerling. Here he had the perfect observation post to see what Nature does when it is practically untouched. Most of what he observed would hardly be visible in the same region at present or elsewhere, so much has the whole of Nature changed since, especially under economic pressure.

Taking on the challenge of harvesting timber from steep slopes in a way that would be financially feasible, Schauberger offered his employers to install a log flume in which he not only replicated but perfected the understanding of the role of water temperature and motion in two ways. In a flume that would follow the course of the river he first attached slats (brake-curves) at an angle within the bed of the flumes such that the water would rotate anticlockwise on left-hand bends and clockwise at right-hand bends. Then he siphoned off old water, which had warmed up, and injected new, colder, water in such a way as to reach everywhere a temperature close to $+4^0C$, the anomaly point at which water is most vibrant, through ingeniously designed valves. Such a technological masterwork paid off and called attention upon him in high places. Already at an early age the Austrian forester demonstrated a key understanding of how water naturally moves and what

---

[48] Viktor Schauberger, Callum Coats, editor, *Nature as Teacher*, 30.

influence is played by movement and subtle variations of temperature, which he would differentiate as positive and negative temperature gradients.

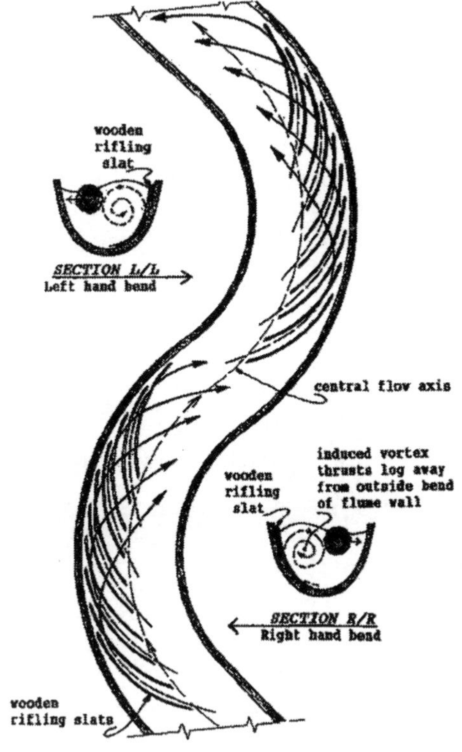

Figure 6: Profile and cross sections of a log flume

After the success of his flume the young forester was named State Consultant for Timber Flotation by the Austrian Federal Minister Buchinger; he was given a salary twice as high as that of other academically trained civil servants in the same field. Academics were doubly frustrated because of the pay differential and because, when they tried to replicate Schauberger's flumes themselves at Reichsraming, Upper Austria, they failed, not having consulted with him, and clearly lacking his understanding. Schauberger was called to correct the mistakes and render the flume operational, after which he built other ones in Tscheppaschlucht and Mürtzal, and yet other places.

In the 1920s Schauberger built nine successful log flumes and renowned hydrologist Professor Philipp Forchheimer was asked to write a report about them by the Austrian government. The meeting with Forchheimer was an important destiny moment. The two were very different and Forchheimer, impressed by Schauberger's achievements, tried to fathom how his colleague operated. He pursued this goal through intense interest, following him like a shadow and asking him question after question. One of these episodes was recorded, which involved knowledge of the temperature gradient in a river and how it behaves. The question/test Schauberger posed his colleague was: 'Can the Professor tell me where the water is coldest, before or after it has flowed around that stone?'[49] Thinking in terms of friction, Forchheimer replied 'warmer' with absolute conviction before Schauberger proceeded, thermometer in hand, to show him that it was actually colder by 0.2°C. Thus challenged, Forchheimer double-checked it only to confirm Schauberger's correctness. This must have left a very vivid impression of the quality of his colleague's observations and knowledge. The collaboration between the two became a friendship and took on humorous overtones. When Schauberger tried to explain why in the log flumes he wanted the water to rotate spirally on its own axis through the application of 'brake-curves', a frustrated Forchheimer replied: 'If you turn me upside down, all that will fall out are formulae. You think in a space that only you and nobody else knows and therefore we cannot make any headway.'[50] This did not fall on deaf ears.

Only after meeting Forchheimer, at age 44, did Schauberger start seriously considering transmitting his ideas. Forchheimer, impressed by the flumes, had asked him to write a paper about them. A further step led to the writing of *Temperature and the Movement of Water*, published in the prestigious Austrian *Journal of Hydrology*. However, this tenuous foothold in academia ceased in 1931 when Forchheimer died.

During most of his life Schauberger was opposed by scientists and technocrats, who were probably mesmerized by what Schauberger was saying in a very different language from theirs, and threatened by his unorthodox, but highly efficient work, not to mention his strong personality. After Forchheimer, a few other scientific patrons stepped forward, chief among them

---

[49] Olof Alexandersson, *Living Water*, 37.
[50] Viktor Schauberger, Callum Coats, editor, *The Water Wizard*, 84.

Professor Werner Zimmermann who encouraged the researcher to write in the magazine *Tau* founded in 1924 in Switzerland. And after Viktor's death his writings appeared in the magazine *Implosion* and *Mensch und Technik*. Zimmermann relentlessly supported Schauberger in his hardest battles for the preservation of the Rhine and the Danube rivers in the years 1935–36, a battle that seemed unfortunately lost in advance.[51] Another collaborator was physics professor Felix Ehrenhaft who helped him with the calculations for his implosion machines, of which more later in Chapter 5.

The ambivalent views of the establishment toward Schauberger are highlighted to the extreme by a bizarre event in the annals of science. In order to safeguard Schauberger's work for the future one of his treatises on 'Turbulence' which looked at the formation and breaking function of vortexes in relation to water temperature, was sealed and preserved at the Austrian Academy of Science under the initiative of Professor Wilhelm Exner, President of the Academy, in 1930. This was done in order to preserve the writing, since at the time it was almost impossible for it to be received favourably, given the prevailing theories on water and the science of hydraulics. The document was unsealed in 1974 and entrusted to the son, Walter.[52]

Schauberger couldn't help but notice the rise of some powerful enemies. Amid growing opposition and manoeuvring of his colleagues he resigned from his post as State Consultant for Timber Flotation. He was then immediately hired to do the same kind of work by a certain Steinhard, leading one of the largest Austrian building operations, to design log flumes in Europe, a work he performed for various years. In 1928 he built a large installation at Neuburg (Western Austria), a project which passed 1400 cubic metres of water in the first hour.[53] The flume operated until 1951 and was described as a 'technical wonder'. Here, as elsewhere, Schauberger faced the disappointment that what was intended to render timberwood more economically profitable was used in ways that exhausted timber resources. But this was a larger problem altogether, one central to our explorations, to which we will return. How the flumes worked can still be seen in the documentary

---

[51] Viktor Schauberger, Callum Coats, editor, *The Water Wizard*, 38.

[52] Callum Coats, *Living Energies*, 5.

[53] Olof Alexandersson, *Living Water*, 29.

Tragendes Wasser, shot in 1930 at the behest of the Austrian Tourist Board.[54]

What Schauberger had devised with the log flumes could be applied directly in the rivers. Where the river had been channelled away from its natural meandering into unnatural straight movements, flow-deflecting guides could be used in order to alternatively guide the water into a clockwise movement alternating with an anticlockwise movement, mimicking in a straight line what naturally happens through meanders from right to left and left to right.

The work with log flumes also branched off in another direction in the twenties. Before 1930 Schauberger also built fourteen of his patented dams. These were designed to address the challenge of cavitation which can cause the slow deterioration of dams, especially those most exposed to western or southern exposures, or hot climates. With knowledge of the effects of the water temperature gradient—of which more will be said in Chapter 3—and simple designs and techniques that he patented, he allowed the concrete to cure completely and effectively in less than a year. He also devised patented solutions to address the problem of water temperature pollution occurring when water is discharged from the lower, colder layers or the top warmer ones of the dams, without consideration of the river's temperature at the point of entry. Professor Forchheimer praised this part of Schauberger's work thus: 'Finally it may be said that Herr Schauberger has already built a number of dams which have proved successful. Some of his structures I myself have inspected, and I can affirm that these new concepts of Shauberger have completely fulfilled the purpose for which they were designed.'[55]

With the builder Steinhard, Schauberger designed and helped create seventeen more flumes, in Bohemia, Czechoslovakia, Hungary, Bulgaria and Romania, most of which remained operational for about twenty years. The timber was easily preserved because of the positive effect of water's natural form of motion. While in Bulgaria he was invited by King Boris to investigate the reasons for decline in agricultural productivity. This led him to recognize the differences caused by materials used in agricultural machinery

---

[54] See Viktor Schauberger Historischer Film *Tragendes Wasser* at https://www.youtube.com/watch?v=QWsUGpKfP8A

[55] Callum Coats, *Living Energies*, 160.

(e.g., wood or steel) and opened a very useful line of inquiry in new directions.

The new research, in collaboration with the engineer Franz Rosenberger, was validated by experiments on how to increase soil fertility, for the most based on choice of materials and forms given to farming implements. A key reasoning was based on the recognition of the role of diamagnetism, among which copper, zinc or silver are prime examples in contrast to ferromagnetism among which we find iron (hence steel), cobalt and nickel. Diamagnetism has a positive effect on levitational forces in the soil or water, whereas ferromagnetism tends to sap both of them of their vitality.

Schauberger recognized how the use of steel ploughs has an effect on the soil analogous to electrolysis, a decomposing effect on water molecules causing a reduction of surface tension—analogous to that of soap or pollutants—and an overall destruction of the soil's subtler energies. Added to this, or rather as a direct consequence of the above, is an increase in pathogenic bacteria, harmful to soil and plants. He further realized the pressure and friction caused by the form of the plough, its angle of penetration and its overall effect on the destruction of the soil capillaries. These were some of the problems he addressed with ploughs of different metal composition and shape, to which we pointed previously.

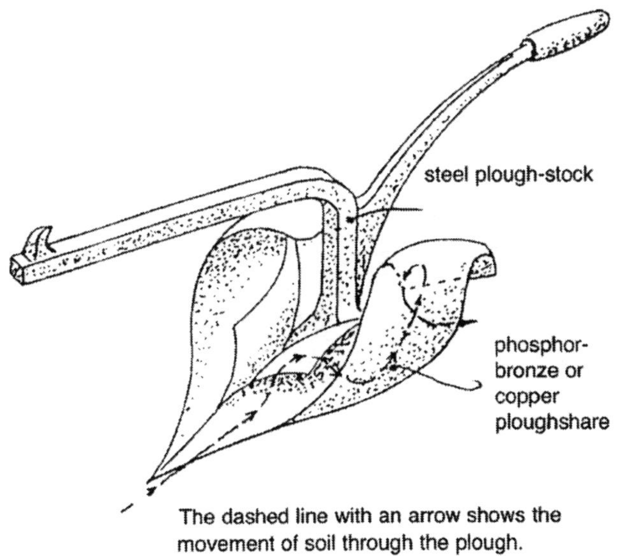

Figure 7: Design of the golden plough

The experiments conducted close to Salzburg in 1948–49 compared the production of wheat and rye in plots ploughed with either steel or copper-plated ploughs. The yield increases of about one third could only be attributed to the use of copper, all other elements being equal. He conducted parallel experiments in Kitzbühel (Tyrol) with potatoes, giving similar results, also noting an overall positive effect on plant health. From here it was a natural step to enter in collaboration with a Hamburg engineer, Jurgen Sauck, in order to produce a copper plough that would emulate the burrowing action of a mole in order to create a long slicing cut before the soil is turned, using a figure of eight movement, rather than the usual ripping motion of conventional ploughs. The new plough is curiously heart-shaped and made of an alloy called phosphor-bronze, just as hard as steel. Around this time Schauberger also experimented with a spiral plough, a hybrid form, halfway between a plough and a harrow, using four or five copper-alloy spiral blades, which were designed in the proximity of Munich. The implement was intended to slice rather than rip, and produce clods, rather than a uniform, pulverized soil.

A research closely related to the above was that on water mains, which put to fruit what the researcher had learned from observing water's movement in rivers and in log flumes. What could not be achieved through a meandering movement in space, Schauberger induced within a straight pipe. Once more it relied on material and movement. For this end he used tight-grained, highly resonant wood and guide-vanes placed at regular intervals in such a way that water would move in a longitudinal, barrelling vortex movement, which cools off and purifies. In essence he did in his pipes something parallel to what he achieved in flumes and rivers. Of this more later.

The researcher's work started branching off in more and more directions, not always easy to follow chronologically. From 1930 to 1933 he researched how to regenerate water up to producing the equivalent of high-quality spring water artificially. Patents were applied for this in 1934. The earliest water-treatment apparatus used a partial vacuum in conjunction with a hyperbolic vortex. It was a first rudimentary prototype, rather bulky and cumbersome. By the early 40s the inventor had designed a far more compact, egg-shaped device. The water he produced was so powerful a remedy that it could help people on the way to recovery from cancer.

Unfortunately it was fought by the medical establishment, and the machine was ultimately confiscated and destroyed.

The work on water naturally verged on energy production since etheric energy was naturally released in the process, and it's not surprising therefore that in the late 30s and early 40s Schauberger developed his 'klimator' a very tight apparatus, the size of a hat, which served as space heater/cooler and generated warm air uniformly through space, a little like a natural Sun radiation. It could also be used as a conditioner. It generated falling and concentrating heat, and when reversed, rising and expanding cold, mimicking the levitational forms of heat and cold that are produced during the day, rather than the reverse common forms that traditional technology produces. We will return to this in the next chapter.

A world threat was looming on the horizon for all Germans and Austrians. In 1934 Schauberger was called in the presence of Hitler, whom he treated with little deference. He spoke to Hitler and his entourage in terms that could hardly be understood and was dismissed. Nevertheless the threat could not easily be laid aside.

During World War II Schauberger was called to serve. Being judged too old he was forced to continue his research for the Third Reich. In 1943 the work was moved to Czechoslovakia, placed under SS supervision and conducted under a high level of secrecy. Schauberger extracted concessions that his co-workers be removed from the camp and allowed to wear civilian clothes. Here he worked at the generation of energy from so-called implosion models, using what amounts to etheric energy.

In 1944 the facilities were moved to Leonstein, Austria. Schauberger at that time worked on some of the things we have mentioned and other ones: a high voltage generator, a machine to biosynthesize hydrogen from water, the suction turbine (also called 'repulsine') or a hydro-electric turbine with corkscrew-like blades. Schauberger believed that other people had successfully tried to replicate what he came up with but, apparently without success.

A vacuum machine, which uses air as its driving medium and very high rotational velocities, was manufactured by the Vienna Kertl company in 1940, and further elaborated at Schloss Schönbrunn under contract for the army. Reports spoke of a flying disc launched in Prague on 19 February, 1945, which rose to an altitude of 15,000 metres in three minutes, cruising at a speed of

2,200 km/hour. Schauberger commented that the machine was supposed to have been destroyed before the end of the war on the orders of a certain Keitel, no doubt an army brass.

At the end of the war the SS shot sixty-two scientists rather than have their secrets passed on to the Allies. Schauberger escaped this fate but ended up being interned by the Americans for a few months and all his equipment in Leonstein was seized. The Russians ransacked his apartment in Vienna.

The research of Schauberger received further academic confirmation in 1952 through experiments conducted at the Stuttgart Technical University. The work was initially sponsored by the Bonn government, eager to discredit the Austrian, a vocal critic of the Rhine river regulation and management. Professor Franz Pöpel, director of the Institute of Hygiene of the university, at first sceptic, was later won over. As we mentioned in the previous chapter, Schauberger studied the conductivity of pipes of glass, copper and spirally-wound copper pipes—similar in shape to the horn of a kudu antelope.[56] Friction was progressively lower in the above order. In the spiral-wound copper pipe not only was friction greatly reduced, it even reached zero and negative values at certain speeds of flow (e.g., at 0.3 litres/sec). The experiments thus invalidated the Second Law of Thermodynamics when water is treated like a living organism.

At this point in his life Schauberger felt a great urge to bring to humanity the most precious part of his legacy, the generation of energy from the ethers. When a British scientist, John Frost, started to work for a Canadian aerospace plant on a flying saucer, he did not go far. The Canadians approached Schauberger with a generous offer to refine the plans but he declined when he could not extract the assurance that the work would be conducted for other than military uses.

Other offers came from Russia, France, Turkey, Italy. The final offer to go to America was preceded by another one by banker and financier Victor Sassoon.[57] In the end Schauberger opted to go to America. The American collaboration did not alleviate his anxieties, quite the contrary. From the documentation available to Callum Coats it seems that the cultural barriers were heightened by mutual incomprehension.

---

[56] Olof Alexandersson, *Living Water*, 118–19.

[57] Viktor Schauberger, Callum Coats, editor, *The Energy Evolution*, 43–44.

The crux of going to America, or taking similar steps, was expressed thus in 1953: '... I have to remain silent, keeping the most pertinent and crucial details from the public until I have managed to find other financiers.'[58] At this point Schauberger's soul balance had been deeply affected. The urge to communicate his work was countered by a deeply wounded man who feared betrayal and exploitation of his work. The added remoteness of the Texas ranch, cultural misunderstanding and commercial concerns became unsurmountable barriers to the continuation of his work and life. Schauberger died on September 25, 1958, a broken man, worried both by the state of the world and the difficulty of continuing his work.

The Phenomenological Approach

Viktor Schauberger looked at aspect after aspect of natural processes and offered solution after solution. However, he did not explain the science behind his work. Even in his relatively simpler flumes people did not learn how to build these themselves. Because of his mistrust of the scientific attitude of his contemporaries he withheld from giving explanations, which meant that people who could have followed him were discouraged. Some of this was at least partly justified in light of some of his more powerful energy-generating technology which required moral responsibility.

So what was it that made Schauberger's science unique? We have already seen how he could retrieve knowledge by letting his consciousness merge with that of the Being at hand—water, trees, plants, animals—and then return and convey what he had learned from the experience, although as we will see this was not an instant process, far from it. That alone would certainly not have been sufficient. Schauberger had the capacity, like Goethe before him, to observe and see what others completely ignored, and to live with questions for a long time, without forcing answers, or escaping into theories. He had maintained his childlike innocence and wonder. However, he knew the science of his time, and could understand it and rebut it. He was also well read in the works of the Greeks, Spinoza and Goethe, among others. Though it's not clear how much he was cognizant of Goethe the scientist, his reverence for Goethe was such that he called him 'prince of poets' and never tired of linking many of his breakthrough insights to lines in his poems. It

---

[58] Viktor Schauberger, Callum Coats, editor, *The Energy Evolution*, 197.

is also clear that he was interested in esoteric literature, to which he probably came close in his collaboration with the magazine *Tau*. We also know that he met Steiner on occasion and had long conversations with him.[59]

Let us now retrace some of Schauberger's steps and listen to his own words. The researcher made it possible to map the course of his growth through his striking observations in the early part of his youth in which he was exposed to marvel after marvel, both in the Schaumburg-Lippe domain and in other pristine areas of Upper Austria. To start with are some examples of questions he posed, excerpted from various sources:

- Why don't wooden posts rot under water, but always above it?
- Why are the banks of south to north flowing watercourses fertile on one side only?
- Why does the heart beat in our breast? Who gives this muscle its impulse to move? Where is the motor for this pump? [The last sentence is said with irony since Schauberger abundantly dismissed the idea that the heart is a pump.]
- How does the trout reach a place above a waterfall of sixty metres?
- Why cold, heavy groundwater high up in the mountains does not sink downwards, and under certain circumstances it even rises?
- Why the more precipitously a clear forest stream flows down a steep slope, the less it damages its bed and banks? And why, when exposed to light [e.g., through deforestation of its banks] does the water become lighter and begin to destroy the banks?
- Why does biomagnetically supercharged riverwater never freeze, and its temperature even increase slightly? And why, when this happens, does watercress begin to grow luxuriantly on the bottom?
- Why is it colder close to the Sun, and the temperature increases with the distance from the Sun?

Many of these questions indicate an uncanny capacity to see, or even sense, things that nobody has seen or bothered about even for decades after the genius' death. Any of these questions consciously pursued to a satisfactory answer, is likely to change one's relationship toward conventional science, which cannot answer them, and rarely seeks to do so.

Much of the wonder of young Schauberger was inspired by popular knowledge which still survived in remote areas of Austria. On

---

[59] Callum Coats, *Living Energies*, vi.

one occasion the forest warden saw a well-known spring covered with a decrepit dome-like structure. Wanting to test the traditional knowledge that a spring has to be sheltered, he had it carefully dismantled, keeping track of each stone. He could thus observe that the spring did dry up in the space of two weeks, as predicted, only to re-emerge when it was covered again with the original structure.[60]

In a spring at Salzkammergut, south-East of Saltzburg, Schauberger observed that in Summer the temperature of the waters hovered around +4°C. In Winter, even at -30°C, the water was 10°C warmer and didn't freeze. During the war the trees around the spring were cleared. The next Spring the waters started to dry up, and the medicinal plants that grew around it gradually disappeared. The forester noticed that shortly after, mange appeared in the area and affected all the mountain goats, unable to feed on the beneficial herbs.[61]

Around springs and waterways the young forester was fascinated by the behaviour of the trout. He first observed how the fish, standing motionless in the centre of the stream, could suddenly dart effortlessly upstream. Even more astoundingly, it was able to completely overcome the pull of gravity. On one occasion, standing by a waterfall, he could observe how the trout entered into a jet of falling water, started 'dancing in a wild spinning movement' and 'floated motionlessly upward'.[62] He further observed that under the right lighting he could actually perceive the path of levitational currents in the waterfall, similar to 'the tunnel in the middle of a circulating vortex of water plunging down a drain', only in this case operating vertically and upward. To achieve this step he noticed that the trout assists itself with strong movement of the gills. In a healthy river the trout swims close to the centre of the stream, in the core-waters which are closest to +4°C, which are also the least turbulent waters. The bodies heavier than water, including the trout's food, travel along this energy line, the true axis of a river. If

---

[60] Alick Bartholomew, *Hidden Nature*, 26.

[61] Olof Alexandersson, *Living Water*, 73–74.

[62] 'With good illumination, the path of the levitational currents is visible now and then. It appears as a tube, apparently empty of water, which we can perceive when falling water circulates with a gurgling sound above a drain [only in the opposite direction]...' (Viktor Schauberger, Callum Coats, editor, *Nature as Teacher*, 86).

temperature conditions change—e.g., through deforestation of the banks, or other man-made causes—the trout will have to seek its food by moving from one side of the river to the other.

Among other, closely related observations, were those already mentioned that under the full Moon or close to it, egg-shaped stones can start gyrating at the bottom of a river, then slowly come up to the surface. The phenomenon is also due to the fact that flowing water in such streams doesn't freeze as long as it is moving; thus it can flow at temperatures of -3 to –5⁰C. It crystallizes on the surfaces of the flowing stones and this helps to bring them to the surface. The egg-shaped stones which could rise were rich in silicon oxide and silicates (of Mg, Ca, Al…); these are present in quartz, rock crystal, flint, granite, sandstone, etc. Once more he was witnessing a magnificently unique phenomenon in which water is activated in such a way as to make it possible to overcome gravity. This observation can be partly reproduced in the setting of the log flumes, which can carry logs of higher specifical weight than that of water, in cold water illuminated by the full Moon. From the wealth of these observations the young forester realized that gravity only operates under certain predetermined conditions, and by changing these it can be overcome.

In the Ödseen (Öd lakes) of Upper Austria Schauberger witnessed a then locally well-known, but little explained, phenomenon. In these lakes, after periods of hot weather the surface of the waters started to draw the dead trees and debris carried by the melting snows and avalanches from the shore into the centre of a growing vortex toward the centre of the lake. The vortex drew them down to the bottom, stripping them of their bark in the process. After the lake returned to apparent calm, another phenomenon followed. Suddenly a roaring sound could be heard from the lake bottom, and waterspouts formed which were as high as a house out of the middle of the waters. They then collapsed and caused ripple waves to flow toward the shore. Schauberger calls this a 'primordial growth of water', a renewal of the water that occurs only in lakes without affluent streams, such as the Ödseen.[63]

Much of the learning that the youth would experience came from venturing in remote, little accessible regions, in the pursuit of some very specific interest. Schauberger, who was a hunter,

---

[63] Viktor Schauberger, Callum Coats, editor, *The Water Wizard*, 35.

ventured in the pursuit of a mountain cock into a part of a primeval forest of predominant spruce and larch on a new Moon night. Bivouacking during the night, while he could hardly see anything else around him, he first recognized a small reddish flame rising from a mound, growing into a fiery egg, with its narrow apex pointing downwards, which then turned into a pale, yellow glow. It kept on growing until it reached six feet high and three feet wide, changing into a magnificent and eerie sight, which even scared him for a while. Placing his wooden staff above it he detected no smell, nor heat. After the formation receded Schauberger saw the ground of the mound covered by delicate flowers of great beauty, which didn't belong to the surroundings, and what was below turned out to be a chamois graveyard. He realized that this phenomenon can only happen when the ground is close to what he called the anomaly point, a temperature of +4°C. These are places which remain at relatively constant, cool temperatures in Summer and Winter, allowing for processes of decay-free decomposition. Such places are naturally sought by the chamois for a comfortable death at any time of the year. Schauberger concluded that '[he was] enriched by an experience which would later help to explain the secret of the transmutation of matter'.[64]

The above and other equally unique experiences, standing separately, remain a puzzle. When they are integrated they each form the threads of the distinct whole that we will explore. At the end of these various experiences Schauberger concluded: 'It was a blinded ox, a shy chamois and finally it was Goethe, the prince of poets, who gave me eyes to see.'[65] This made him adopt the motto '*Kapieren und Kopieren*'—understand and copy Nature—which he would not tire to repeat.

---

[64] Viktor Schauberger, Callum Coats, editor, *Nature as Teacher*, 48.

[65] Viktor Schauberger, Callum Coats, editor, *The Water Wizard*, 46. The reference to the blinded ox is one that contributes to the other experiences, but in negative. It refers to the episode of seeing an ox collapse in the attempt to carry heavy timber down the mountain on a log-sledge, whereas a log flume could have done it without energy and the need for such a sacrifice (Viktor Schauberger, Callum Coats, editor, *The Water Wizard*, 40–41). The episode of the shy chamois had to do with the fact that previous to the discovery of the dancing logs and stones, Schauberger had hunted down a chamois but it had fallen way down into the ravine in which he witnessed the phenomena of the rising stones. (Callum Coats, *Living Energies*, 140.)

We will finish this part of the exploration of Schauberger's scientific methodology by turning to further thoughts and to his own words. Very revealing statements as the following were repeated by Schauberger in many ways: 'In Nature everything happens indirectly and it is therefore not enough merely to think logically. Whoever desires to evolve personally or to contribute towards the general improvement, will have to get used to thinking and acting biologically.'[66] We will see examples of what he means in the working of the temperature gradient in water, so central to water's motion and the forming of its banks, or in the discernment of the process of growth in the forest.

Schauberger in essence cautioned us to believe that Nature's secrets are revealed on the surface of things. 'Nature is careful and operates with real values, and the more valuable they are, the less they manifest themselves.'[67] Her secrets cannot be found through simple cleverness. Real interest and a devoted heart must precede the growth of insights. The examples given in relation to Schauberger's inner growth are reminiscent of how ideas matured in J. W. Goethe the scientist. To patient observation and the maturation of questions over the space of years, followed insights that could stand the test of time. Along these lines Schauberger challenges us: 'The majority believes that everything hard to comprehend must be very profound. This is incorrect. What is hard to understand is what is immature, unclear and often false. The highest wisdom is simple and passes through the brain directly into the heart.'[68] For Schauberger Nature is endowed with intelligence and meaning and an evolutionary thrust to generate ever more complex life-forms and higher levels of consciousness. The human being is truly the crown of creation, but this is not meant in an old religious sense. This was something he had long intuited.

Finally, as any true Goethean or spiritual scientist fully knows, the crux of science rests in the link between correct observation and correct interpretation and in the role that our intellect plays in between the two. Is our given, intellectual thinking sufficient to give us a reading of the facts, or do we need to transform it? Schauberger predicated in a general way that science needs to think 'an

---

[66] Viktor Schauberger, Callum Coats, editor, *The Fertile Earth*, 156.

[67] Viktor Schauberger, Callum Coats, editor, *The Energy Evolution: Harnessing Free Energy from Nature*, 98.

[68] Callum Coats, *Living Energies*, 28.

octave higher'. He realized here lay the difficulty in communicating his knowledge to the younger generations, and in this effort he probably fell short.

Encapsulating his thoughts about the dangers and limits of present human thinking he stated: 'Everything in Nature is built up sequentially, step by step. Therefore as a rule the intellect is something of an outlaw. It runs the gauntlet of a thousand dangers and is subjected to perils more hazardous than unsuspecting humans can imagine.' From here he goes on to criticize Kant: 'Kant made the fallacious assertion that any knowledge to which mathematics cannot be applied cannot be described as science. Instead of becoming a creator on earth, humanity became a vulgar speculator, more and more estranged from truth and reality.'[69] The researcher realized that this is a challenging task: 'Only those with exceptional intuitive abilities, and thus of an artistic turn of mind, will actually be able to grapple with this extremely difficult way of thinking.'[70] With this last statement everything comes full circle and we can understand Schauberger's love for Goethe. Like his predecessor he was predicating the alliance of the scientific outlook with an artistic mindset. Many of Schauberger's descriptions of natural phenomena and the way in which they opened a door of understanding are in fact delightful, miniature artistic renderings, worth reading for sheer delight on top of the revelations they offer. Two short ones are offered at the end of this chapter to offer us a feeling for how Schauberger thought and wrote.

---

[69] Viktor Schauberger, Callum Coats, editor, *Nature as Teacher*, 30. Further on this theme, 'Our thinking is inconsistent with what we actually see. ... Our sight constitutes an unconscious, automatic transformation process, whereby the negative image—like a photographic negative—i.e, the effect, is transformed into a positive one, like a diapositive color slide. Our thinking, however, is really a purely individual, conscious process and therefore learnable. If our thinking is to attain the same perfection as our seeing, then we must change our way of thinking and learn to see reality, not as an action, but as a reaction. Perfect thought lies in the apprehension of the correct reaction, for before the eye can show us the positive, it must first transform the negative and in a certain manner must break up what it records. What we see, therefore, is the turning inside out of what we receive. What our mind grasps in this way must be reformed and rethought if we wish to attain that for which we strive.' (Callum Coats, *Living Energies*, 6.)

[70] Viktor Schauberger, Callum Coats, editor, *Nature as Teacher*, 34.

Schauberger was an adept of both the artistic mindset and scientific outlook. He naturally used his intuition through his unique ability to observe carefully. He was then able to understand through concepts, and apply these to his inventions. Synthesis was for him akin to natural, planetary motion—related to the forces of levity—analysis to what he called technical motion, related to gravity. He further recognized that these two processes must constantly play out in equilibrium, at the theoretical and practical levels.

Schauberger was further aware that he was introducing new concepts for which there was no adequate lexicon. 'Today this train of thought would appear utopian. All the more so, because it is exceedingly difficult to differentiate contemporary concepts from those required here, for which there are no technical terms. Therefore modern and frequently inappropriate terminology will have to be used. By placing them between quotation marks a different sense or meaning is intended.'[71] An example is his use of the word/particle *ur*, which the reader often comes across, to signify primeval, primordial, primal, first principle, something pertaining or belonging to, or existing from the earliest beginnings.

It was overall a pity that beyond occasional meetings a full connection did not occur between Schauberger and Rudolf Steiner, through which the forester could have found access to the objective language and concepts he most needed. There were other attempts here and there that have been recorded. In 1955 Schauberger received a book entitled *Fundamental Principles of Potentization Research* provided by Lilly Kolisko, no doubt to show the connection between the work of both scientists in understanding that in Nature the movement from the physical to the energetic/etheric levels happens all the time, and that it can be replicated and used in the laboratory, in this case through potentization. Schauberger, then more pointedly focused on the matter of energy production, recorded: 'It was to no avail.'[72]

This failed connection is all the more to be decried since Schauberger clearly had an interest in and understanding of esoteric matters: witness, if nothing else, his comments and insights on Atlantis. He knew that in the lost continent the high priests had abused the

---

[71] Viktor Schauberger, Callum Coats, editor, *Nature as Teacher*, 154.
[72] Viktor Schauberger, Callum Coats, editor, *The Energy Evolution*, 198.

etheric forces.[73] Had Schauberger fully made contact with spiritual science he could have double-checked the results of spiritual vision and avoided problems of interpretation of his perceptions that crop up here and there in his work.

Ingeborg Schauberger, the Austrian's daughter-in-law, helps us assess Schauberger's unique personality and some of his shortcomings. From her we understand that he was a very eloquent speaker. People understood him in the moment but could not rebuild the lines of thought soon afterwards. 'They remembered the feeling, not the understanding.'[74] Even his devoted followers could not fully understand him. He was impatient and tended to outbursts of anger. His responsibility weighed on him with a feeling of being unable to bring out everything he carried inside. He tended to blame everybody else other than himself for their lack of understanding. He even distrusted his son, Walter, only relenting late in life to let him in on his secrets. He finally understood that his son could speak to the scientists, in a way he couldn't.

The son, Walter, trained as a physicist, and by the fifties he started to appreciate his father's work, giving up his academic career for this reason. In 1949 father and son were part of starting what was likely the first green political party, the Green Front.[75] Walter soon realized that the answers would have to come from outside politics, and he set out to restore his father's legacy and reputation. He had to come up with yet newer scientific terminology to explain the work of his father.

Schauberger on Bio-Energies

We close the chapter with an example of Schauberger's discoveries, which offers us a taste for the way in which the forester expressed himself. In the first he speaks of triboluminescence, in the second of

---

[73] 'It was therefore inevitable that we should be assailed by the catastrophes of Nature that we are presently experiencing. They are exactly the opposite of those experienced by the Atlanteans, who realized the danger of over-stimulating formative and levitative energies only when whole sections of the Earth were wrenched skywards by these elemental (metaphysical) suctional forces.' From Viktor Schauberger, Callum Coats, editor, *The Energy Evolution*, 79.

[74] Jane Cobbald, *Viktor Schauberger*, 152.

[75] Jane Cobbald, *Viktor Schauberger: A Life of Learning from Nature*, 135.

peculiar phenomena showing the power of etheric energy at cold temperatures, under the effect of a full Moon. The two phenomena he describes are interrelated.

Triboluminescence[76]

Triboluminescence is the natural creation of electricity at the bottom of a pristine river. This is an energy which does not interfere with electrical devices; in effect a type of bioelectricity.

> The present dirty grey, muddy brew known as the Blue Danube, upon whose bed river-gold once gleamed, and the Rhine, the symbol of German identity, where Rhinegold flashed in bygone days, are tragic testimonials to these perverse practices [of modern river regulation]. The mythical gold of the Nibelungs originated in the golden glow given off by pebbles as they rubbed against each other while rolling down along the riverbed at night, for when there is a decrease in water temperature, the tractive force [related to the etheric energy] increases, causing the stones to move. If two pebbles are rubbed together underwater, a golden glow appears. This yellowish-red fiery glow used to be mistaken for the flashing of gold, thought to be lying on the bottom of the river. Today this 'river-gold lies' heaped up in huge amounts of gravel, shifted hither and thither by the force of the sluggish and murky water masses flowing above them. They no longer imbue the water with *energy* and *soul*, as they once did. Instead they assist in ousting the soul-less body—water—from its badly regulated course.

Dancing Logs and Stones

The following excerpt is taken from an article that appeared in *Implosion* magazine # 7.[77]

> On a particularly cold December morning after the above event had occurred, I was stalking a very powerful buck in the ravine where the logs always came to rest. ….
>
> Suddenly I slipped and tumbled into the ravine down the icy path of an avalanche. Using my staff as a break, I landed luckily on a heap of snow. … At this point the water was several metres deep, crystal clear and totally still. While washing my sweaty hands, I noticed

---

[76] Viktor Schauberger, Callum Coats, editor, *The Water Wizard*, 16–17.
[77] Quoted in Viktor Schauberger, *Nature as Teacher*, 43–45.

many sinkers about four metres (13′) below. They were performing a remarkable dance. It was as though the logs lying in the absolutely still water had become magnetic. Now and then the butt of one log lifted, floated upwards slightly, laying itself a few feet under another log, only to recoil in the opposite direction the very next moment as if in fright. Somewhere, horns and beard [of the chamois] were forgotten! Hour upon hour passed. I was unable to drag myself away from this extraordinary spectacle. Gradually evening fell and with it came an even keener chill. Suddenly a heavy log stood straight up, paused for a moment, and with a lurch shot out of the water, to be encircled immediately with an Elizabethan-style ruff of ice. Soon the other logs began to perform the same witches' dance. For a few minutes several logs, girdled with a nice necklace and projecting about 10 cm (4") above the surface, swayed to and fro in the water as if alive. Indeed, even more marvelous events were about to take place! For a long time I lay on the ice and observed the action at the bottom of the icy-cold stream. There! I couldn't believe my eyes! There was a strange activity amongst the variously-sized stones lying on the bed. A few were as large as a human head and began to come to life. For a long time they played the same as the logs did earlier, first approaching each other only immediately to recoil. Flouting all the laws of gravity they drifted about, attracting and repelling each other. At first I thought that the stones might be electrically charged. I remember the phenomenal luminescence given off by such milky-colored stones which leave yellowish-gold comet tails behind them when rubbed together underwater. This phenomenon is what had apparently given rise to the legendary Rhinegold in the 'Song of the Nibelungs'.

All of a sudden a large, head-sized stone began to gyrate slowly in circles like the trout at the waterfall before it begins to float upwards. It was egg-shaped. The very next moment the stone was at the surface right before my very eyes. It was quickly encircled by a visibly growing collar of ice and this apparently-possessed, gently rocking stone, floated in the full moonlit surface of the water, just as the heavy beech logs had done earlier on. Soon a second, then a third, and still more stones, all making the same maneuvers, rose to the surface of the water. The water soon looked like a cake bespeckled with raisins. Eventually almost all the milky-white stones that had been smoothed and rounded rose to the surface. The remaining rough, angular stones which had fallen in from the banks were left motionless on the bed of the stream.

At the time I took no account of the precise angle of the incident moonbeams, which make lunatics walk about on roofs. Naturally I had no idea that this involved a concentrative process, nor did I realize the significance of the oxygen-concentrating effects of the chill on this bitterly cold night of the full Moon. This cold concentration gives rise to the emission of expansive emanations, which lead to the creation of the 'original' form of motion. It is this form of movement that overcomes gravity and raises the specifically heavier stones to the surface of the water. As a huntsman I was well aware that female physical masses, charged with negative ions, became fiery when handled coolly. I also knew that in their hunger for reactive substances (semen or fertilization), these female bodies, known to triumph over the male when they succumb, were able to overcome not only their own weight, but also that of their superimposed load. What I found hard to believe was that, apart from negating their own weight, these stones were also able to overpower the watery resistance to motion pressing down upon them.

Although at the time I was unaware that this process was the starting point for the explanation of atomic transformation, it nevertheless made me decide to be even more observant in the future. As a result further events unfolded which, understandably, appear highly mystical to scientists, but which in reality are completely natural manifestations of the forces of Nature.

We find in this example many of the elements which will accompany us in our further explorations: the power of etheric forces in primeval, pristine environments, the influences of original or planetary motion, the role of temperature and of form among others.

Chapter 3

A Holistic Understanding of Water and Rivers

'Naturally-moving water multiplies itself. It raises its own quality and wells up autonomously. It changes its freezing and boiling points. Wise nature makes use of this phenomenon to raise water to the highest mountain peaks without pumps...'[78]

Viktor Schauberger

Water has been a never-ending source of wonder and inquiry for decades of my life. It started with flowforms and the work of John Wilkes, when I studied at Emerson College in 1982–83. It continued with the work of Rudolf Hauschka and the understanding of how water renders possible potentizing of homeopathic medicines. It picked up pace with the work of Theodor Schwenk and son Wolfram, and of Masaru Emoto, … . It found practical applications that I explored in the work of Johan Grander, or Clayton Nolte and devices I could use in the house to improve water quality. In between it led me to the work of Viktor Schauberger, who has rightly been called the 'water wizard.'

We can approach an understanding of water from a multitude of aspects, but one of them stands out among the others for its simplicity and for the richness of insights it can provide for the rest of our explorations. It will contribute to 'raising our thinking up an octave' in immediately practical ways.

The Vortex: A Unique Phenomenon

Imagine a vortex like you may have seen in countless images of water circling toward a centre, or of galaxies delineating a centre point. Here we have the counterpart to centrifugal forces, which are easily comprehensible from our immediate environment: take a clothes washer or any compression engine. They all use centrifugal forces. A combustion engine operates with tremendous forces at the centre, which then move to the periphery. Energy is dissipated from a centre outwards. In the vortex we have the polar opposite to this phenomenon, the natural complement: forces converge from the

---

[78] Viktor Schauberger, Callum Coats, editor, *Nature as Teacher*, 108.

periphery inward. The movement of the vortex mirrors that of the planetary system with the Sun at its centre. The speed of movement in this case accelerates towards the centre. And planets circle faster the closer they are to the Sun, such as Venus or Mercury, slower the farthest away, such as Saturn or Pluto.

Water that drains naturally displays the galaxy-like configuration of the vortex, anticlockwise in northern hemisphere, clockwise in the southern. Looking up closer we will notice that the whole system sways and moves. A vortex has a natural pulsation; it shrinks or expands in diameter, with corresponding extension and contraction of its length, all of this happening in a periodic manner.

Let us look closely at a vortex generated in a cylindrical glass vessel. At the edge of the vessel we have a very slow rotation; moving in the speed increases dramatically. At the centre and at the tip of the vortex the speed is virtually infinite. In physics the phenomenon is described by the equation $r \times v = c$, with r being the radius, the distance of the surface of water from the centre of the vortex, c is a constant number and v the speed of rotation (e. g., cycles per minute). From this equation we see that when $r = 0$ then $v = \infty$. At the tip of the vortex the tendency toward infinite speed is only countered by friction. Since this speed cannot occur in the physical world, what happens is that water begins to dissociate into vapour, releasing electrical charge.

When we investigate pressure in a vortex we find something similar to the velocity. When $r = 0$, then $p = \infty$. In physics this means a pressure lower than a vacuum. We have here described the polar opposite of a combustion engine at whose centre occurs an explosion radiating outwards. Here the speed and amount of energy moves from the periphery inwards, approaching a point of infinity within. This is what Schauberger called a biological vacuum. The energy generated he called implosion to contrast it with the explosion that takes place in a combustion engine.

Under the conditions we find at the tip of the vortex there is nothing else than surface tension This tension is like an internal pressure: it increases with the decreasing size of a droplet. The pressure difference becomes enormous when moving toward the molecular size. Practically speaking, the fluid reaches a state in which it is torn apart. 'Physical conditions in this centre surpass the loading capacity of material substance beyond all measure. The most attenuated ethereal medium cannot withstand such an immeasurably great suction, such abnormal stress. Physical substance bursts asunder under such impossible conditions,' comments physicist Georg Blattmann.[79] We are reaching a boundary between two orders of reality, one visible and well-known, the other no less real, though only understandable by a qualitatively different kind of thinking than Euclidian geometry and ordinary mathematics.

The scientist Patrick Flanagan perfected a laboratory arrangement in which he used a clear egg-shaped ellipsoid, a nearly cylindrical vessel, with a hole pierced through the bottom. The rhythmic pulsations and movements of the layers of water moving at different speeds were made visible through drops of food colouring. When the speed of rotation is increased the diameter of the vortex throat shrinks. We saw that as the diameter approaches zero, the velocity approaches infinity. Flanagan carefully measured the electrical charge of the vortex with a wire electrode set into the centre of the throat—but without touching water—and another one in contact with the water. At 1000 rpms he recorded an electrical potential of more than 10,000 volts.[80] He also tested the systems by reversing the direction of rotation, concluding 'Steiner's idea that an energy enters the water each time the direction of stirring is reversed [e.g., in the stirring of biodynamic preparations] is right on the mark.' Looking specifically at what happens when the main vortexes collapse led him to the observation that the water is then filled with millions, if not billions of minute vortexes. Relating to the rhythmical stirring of biodynamic preparations he concluded that the energy 'has to be absorbed into the hydrogen bonds of the water and be absorbed into the particles of the 500, rendering them colloidal, and readily ingestible by both the microorganisms and the hungry single-celled root hairs of plants.'[81]

The conditions within a vortex—immense speeds, immense suction, fluid torn apart, potential of more than 10,000 volts, millions of minute vortexes—form the preconditions for what Schauberger calls a 'biological vacuum'. This is what he put to use for the

---

[79] Georg Blattmann, *The Sun: The Ancient Mysteries and a New Physics*, 71.

[80] Peter Tompkins and Christopher Bird, *Secrets of the Soil: A Fascinating Account of Recent Breakthroughs—Scientific and Spiritual—That Can Save Your Garden or Farm*, 106–09.

[81] Ibid., 113. Flanagan is here referring to the stirring of the biodynamic preparation horn manure, or 500, which is stirred clockwise and anticlockwise in order to form vortexes in rhythmical alternation.

generation of bioenergies. But before getting to these, we'll see the vortex play a primordial role in the river's health.

Thinking in Polarities

We have just witnessed in the vortex a term that forms a contrast with most common experiences that are offered to our senses. We are familiar with, and can understand intellectually, the phenomena of gravity, the working of explosion engines, or everything that has to do with weight, measure and number. The phenomenon of the vortex forms a perfect counterweight to that of the waterfall. The latter can be explained by the commonly known physical laws mentioned above; the former asks us to make recourse to a qualitatively different kind of thinking.

Understanding Nature means reconciling apparent contradictions or mutually limiting antitheses; life and death, wakefulness and sleep, light and darkness, expansion and contraction, lightness and heaviness, etc. In Nature these opposites work in and through each other; they are the terms of a polarity. They are like parts of a continuum, where one of the two, though predominant, always works with the other, even when we move toward the end of the continuum. At the centre of these contrasts lies the question of quality, which will lead us to a breakthrough in the understanding of Nature.

Why is quality important? To the accurate observer it is clear that nothing is ever the same/identical in Nature. Here always appear the greatest change and heterogeneity. Nothing can be gained for evolutionary purposes from sameness and repetition. Life is created out of differences, however minute and subtle.

Modern scientific observation knows Nature in its quantitative aspects. No qualitative aspects enter the equation unless they can be first defined and then turned into a quantitative analysis; but then we are back to square one. This limitation of science to the purely quantitative rests on the exclusive use of indirect observation/measurement through technological means (chemical analysis, spectroscopes, X-rays, Geiger counter, electrical measurements, etc.) or reliance on passive sense experience—experiences which are not strengthened through deeper integration and phenomena-based thinking, as one does in Goethean science.

Qualities cannot be understood without an active sense experience. They remain occult. In the Goethean approach, the terms of polarity can be reconciled by bringing together the analytical element of thought with the feeling element in art, as Schauberger himself intuited. The rigorous scientific analysis can be strengthened by the no less rigorous, but subtle, living feeling/artistic perception. This is how Goethe brought science to a new level. Through him passive thinking became active, living thinking. He achieved what Rudolf Steiner called 'Imagination', a sense-free perception of the archetypes behind physical reality. Schauberger worked very much in the same direction. Like Goethe he left us the results that we can now begin to understand through a spiritual-scientific approach, though needing to overcome the difficult problem of the author's language, especially for those of us who do not have direct spiritual perception.

Goethe's work led, among other things, to the concept/experience of polarity, which is central to the sampling of phenomena we have already announced and to the deeper exploration that will follow. In Goethe's work these were the polarities of horizontal/vertical, light and darkness, Earth and Sun, expansion and contraction, etc. In polarities we can say that quantity is the thesis and quality the antithesis. However, quality is the determining factor. Quality is the differentiator and animator of life. An example of this: levity is what allows the phenomenon of gravity; the latter is the force of resistance. Gravity is present within restricted conditions within our solar system, levity is much more pervasive.

In commenting on Schauberger's work Callum Coats indicates that the polarities of cold and heat, suction and pressure, explosion and implosion, centripetence and centrifugence are the antitheses and 'the agencies of self-organizing, intermediate, vibratory matrices of immaterial energies by which the gap between the Will-to-create and creation, spirit and matter, and idea and manifestation is bridged'.[82] The full attainment of one term of the polarity would fully negate the reality of the other term, but also negate itself because it is only justified in a relationship. In the timeless dimension polarity corresponds to two sides of a single power. The higher unity outside of time becomes opposition in the realm of duality. At the purely intellectual level polarity is a paradox; at the level of experience it is lived as a mystery. Through the attainment of Imagination, and further forms of cognition, it can be both understood and experienced.

---

[82] Callum Coats, *Living Energies*, 74.

At the human level polarity embraces the core of our being and deeper experience of life itself. It is through the existential agency of polarity that each one of us can experience being truly individualized, while at the same time truly universal. In fact the more individualized we become, the more universal we can feel. This is a primeval, existential kind of polarity.

As part of Schauberger's 'thinking in higher octaves', and central to our understanding of everything that will follow is the encompassing of polarities:

Matter	Spirit
Heat	Cold
Gravity	Levity
Pressure	Suction
Explosion	Implosion
Centrifugence	Centripetence
Oxygen(es) (Sun)	Carbones (Earth)

An example that will be central to our understanding of water: the polarity of cold and heat. Currently there is only one view about the polarity of heat and cold: heat rises and expands and cold sinks and contracts. And this is valid within all technological systems. But Nature also uses the opposites of rising and expanding cold and falling and concentrating heat.

A very common example will suffice. During daytime, if we go from the top to a bottom of a valley we experience increasing warmth (falling and concentrating heat); when we ascend it becomes progressively cooler (rising and expanding cold). At night the process is reversed; as we descend the air is chillier and denser (falling and concentrating cold) and as we ascend the air warms (rising and expanding heat). The rising and expanding cold and falling and concentrating heat is what makes life possible, and they must prevail over their counterparts. They have a life-affirming function, leading to what we can call 'cold, formative, metabolic processes'. The reverse—falling and concentrating cold and rising and expanding heat—is what Nature uses for organic decomposition, for decay without putrefaction. At present these are the processes exclusively used and amplified in modern technology.

Continuing the work of his father, Walter Schauberger indicates: 'The manifestation of all natural forces is the result of the interaction between two opposites, neither of which ever reaches totality in the lower realm of duality (the physical world), for they can only become total when they unite within their unifying, nonphysical, governing principle. In the physical world each component of a pair of forces can only attain 96% of its boundary or extreme condition. Once this point is reached, then its opposite force gradually begins to gain strength.'[83] An example: in a vacuum there is always a 4% left of the medium to be evacuated.

Armed with the understanding of polarity we will ask questions that modern science doesn't ask even when it meets with outer phenomena that defy simple understanding. Most of the time these, it seems, cannot even be recognized. They were seen, however, by Goethe, Steiner, Hauschka, Schauberger and many others.

A Fuller Understanding of Water

'In Nature all life is a question of the minutest, but extremely precisely graduated differences in the particular thermal motion within every single body, which continuously changes in rhythm with the processes of pulsation.'[84]

Viktor Schauberger

If Schauberger's motto: 'Good water—good life! Bad water—bad life! No water—no life!' may sound simplistic at this point, I invite you to revisit it at the end of this or the next chapter. Upon the very precise premise of good water, Schauberger envisioned how to transport timber and materials down steep slope gradients through log flumes; regulate watercourses without embankment works; control the river flow through the temperature gradient and the help of flow-deflecting guides; raise the height of the water table around the rivers; produce high-grade drinking water; render dams more long-lasting and avoid the problems they generate in their water discharge due to temperature differentials; raise water without pumping devices, etc.

Water's Unique Properties

Schauberger reasons cogently that since without death there would be no life, there must be two forms of motion. There is on one hand a motion that produces/maintains what is alive, and on the other a condition that sets the stage/preconditions for this to happen from what is decayed, and these are two completely different

---

[83] Callum Coats, *Living Energies*, 62.

[84] Viktor Schauberger, Callum Coats, editor, *The Water Wizard*, 6.

kinds of motion. On one side what we will recognize as planetary motion—to which our exploration of the vortex has already pointed—a motion that is the prerequisite for everything that manifests life, and on the other a motion that brings about the formation of those products through which the first one can operate. This is in fact the kernel of our present exploration. But before that we will turn to water's unique properties.

Water behaves in ways hardly conceivable from general chemical theory. In comparison with similar substances one would expect its freezing point to be at –120°C (-184°F), its boiling point at –75°C (-103°F).[85] That this is not so is what renders life on Earth possible. Water also has the highest surface tension among comparable substances, another of its important traits. This is what allows a dewdrop to maintain its coherence. Water also has a great dielectric value. A vacuum has a dielectric value of 1; pure water of 81. It's almost the highest dielectric value in Nature, conferring a great innate resistance to the transfer of charge. Water thus allows for large electric fields in living cells as we have seen in the case of the vortex. Not surprisingly water also has a great specific heat, which means great inertia to changes of temperature. Given a certain amount of heat, the temperature of water rises more slowly than that of similar materials.

Water dissolves oxygen, nitrogen, carbon dioxide, from the air, calcium, potassium, sodium, manganese, and many other elements from the rocks. It combines in fact with more substances than any other molecule and is called for this reason the 'universal solvent'. 'Both water and our bodies contain 84 elements [of 103 known elements] in the same proportion.'[86] Water collects all these substances and deposits them for new growth.

Water sacrifices itself completely to the environment, for good or for bad. In its circulation through atmosphere and soil it will evolve from an immature state—generally associated with little substance in solution and gases—to a mature state in which it accumulates dissolved solid substances and gases. When it is healthy it pulsates like blood, it spirals and vibrates, and this in turn maintains its vitality. To quite a degree it purifies. However, when it is immature, or degraded/polluted, it greedily takes up minerals and nutrients. When immature it will take out minerals and trace elements out of our bodies.

---

[85] Jochen Schwuchow, John Wilkes, Iain Trousdell, *Energizing Water*, 10.
[86] Alick Bartholomew, *Hidden Nature*, 109.

The movement generated by the vortexes in flowing water creates micro-clusters and a complex laminar structure which can be observed under a microscope. The structures are called 'crystalline-fluid' because they display a degree of order almost as high as that of a crystal. The clusters can store vibrational impressions or 'imprints'. Higher order clusters water vibrates at high frequencies.[87] At +37°C water can form clusters of 300 to 400 molecules; the colder the temperature the longer the clusters will become.[88] This structure probably increases as we get to the boundary temperature of +4°C, with the approaching of the full Moon, etc.

The greater the activity of vortexes the more water can store information. With beneficial imprints water health can be restored a little bit like through homeopathy. The opposite is true of degraded, damaged water in the negative direction. We know through the work of Masaru Emoto that water considered legally potable may not be able to crystallize, meaning it has lost vitality and still carries the information of the initial pollutants. Safe to drink and good to drink are two worlds apart! This is because water can maintain the information of substances it has dissolved, even if they are no longer there.

Harmful substances have specific frequencies, and water, it seems, absorbs them as soon as it enters in contact with them, leading Schwuchow and others to conclude: 'The photon (quantum light) spectrum of contaminated water thus differs very significantly from that of clean water.'[89] And even after physico-chemical treatment these frequencies continue to be harmful for human use. Not only that: water thus processed most often loses its rhythm/movement properties.

In order to erase the information imprinted on water we can heat it at 400°C, or submit it to vigorous vortexing. This is what happens in homeopathy or in flowforms, in both instances through specific kinds of motion. This is also what is done in a growing variety of water devices, which first create chaos (vortexing), then eventually re-imprint new information into the flowing water. An example is the Grander water device, which uses high frequency water, encased in a devise to reprogramme water after submitting

---

[87] Jochen Schwuchow, John Wilkes, Iain Trousdell, *Energizing Water*, 83.
[88] Jochen Schwuchow, John Wilkes, Iain Trousdell, *Energizing Water*, 10.
[89] Jochen Schwuchow, John Wilkes, Iain Trousdell, *Energizing Water*, 14.

it to vortexing. The two waters do not enter into physical contact.[90] On a much larger scale Schauberger tried to restore water's health through reintroducing natural movement in the riverbed.

New Findings

Schauberger applied much of his research to understanding the nature of water through a variety of complementary and overlapping approaches. Water has a cycle in which it goes from the atmosphere to the underground and back again. Forests and rivers, have a tremendous impact on this cycle. Basically, according to how rivers and forests are managed we have two very different global outcomes. All of this can only be understood if we first address water's qualities, those that are created by the interaction of temperature and motion. We will return to the ultimate aspects of the question, and how the water cycle is affected at the end of the next chapter. This is because this is part of a complex understanding that needs to be built up patiently.

A primary aspect of the cycle of water is the one involving the integration of chemical compounds from atmosphere and geosphere. Water is continuously weaving between heights and depths, and can only reach an optimal state of balance and health through the integration of these. Sun and Earth and their influences form a polarity in the growth of that living organism which is water. High quality, mature water contains a balance of geospheric and atmospheric elements and energies. In accordance with much of traditional knowledge Schauberger attributed to the first a feminine quality, to the second ones a masculine quality.

To understand the maturation of water we will look at the contrast between 'carbones' and oxygen. What Coats translated as carbones is what Schauberger understood as '*Kohle-stoffe*' or 'mother substances'.[91] Under the term carbones are grouped all compounds of carbon, dissolved salts and all other mineral and metal

---

[90] Hans Kronberger and Siegbert Lattacher, *On the Track of Water's Secrets: from Viktor Schauberger to Johannes Grander*; for an understanding of Grander's water device see Chapter 4. Other ones known to the author are the devices of Natural Action Technologies developed by Clayton Nolte or the the Living Water Vortex jug developed by Clean Water in Denmark. This is to say that this is a growing field and what is shared here is part of a larger whole.

[91] Viktor Schauberger, Callum Coats, editor, *Nature as Teacher*, 17.

compounds, all except oxygen and hydrogen. The term acquires a fuller meaning when we recognize the importance of carbon-rich substances in coal and oil reserves which Schauberger saw as the source of carbones for the production of a good, healthy water, rich in carbonic acid as the water rises from the depths. Carbones and oxygen play a complementary role in relation to temperature and motion which will be critical to water's behaviour. Before going into details we can indicate that when one is active the other becomes passive and vice versa.

On their ascending movement through the soil, water rich in carbones generates subtle etheric energy, levitational forces overcoming the forces of gravity. They seek to be fertilized/are attracted by the descending oxygen and their male, fine energies counterpart.

Mature water acquires its highest levitative force when it rises through the layers of the Earth: the longer the path the water travels the riper it will become. The most mature water is spring water, which contains a large amount of dissolved salts and a high quota of free and bound gases (e.g., CO_2 and O_2).[92] At the other end of the spectrum lies rainwater, a form of juvenile water, far from the state of equilibrium of mature water, with a high oxygen content, little or no dissolved carbon dioxide or carbonic acid (H_2CO_3), and little or no salts.

When the downward movement of oxygen intersects with the upward movement of carbones, energy is made available. 'Through the resistance arising from the interactions between carbone and oxygen, fluctuations in temperature again occur and with them the impulse to move—the pulsation of water, which in this way at times dissolves salts and at others deposits them, transports them, creates energies and transforms them.'[93] To understand the pulsation of water consider that water moving toward the anomaly point of $+4^0C$ breathes in—acquires more gases, condenses. Water moving away from the anomaly point deposits gases and substances and expands; it breathes out. At $+4\ ^0C$ water has a host of unique characteristics to which we will return; this is the temperature at which it acquires its highest density and the one in which it can dissolve the highest amount of salts and gases. For this reason it has been called the anomaly point.

---

[92] Viktor Schauberger, Callum Coats, editor, *The Water Wizard*, 72.

[93] Viktor Schauberger, Callum Coats, editor, *The Energy Evolution*, 7.

The utmost difference between kinds of water is their behaviour in the ground and in the human body. In moving from the depths to the surface, water changes from being a 'taker' into being a 'giver'. It either absorbs from the ground or dissolves. At the end of its maturation water can deliver the widest variety of dissolved compounds and ionized elements in homeopathic doses to the living systems of its environment. In the zone of the roots the associated microorganisms, living in symbiosis, transform the carried substances into larger molecules which are transported by the capillaries of the roots.

Juvenile and Mature Waters

Let us look at the different kinds of water, from rainwater to spring water. It is well-known that distilled water is not fit for human consumption. It is brought out here because it highlights much of what follows. Distilled water is the most extreme example of 'synthetic juvenile-like' water. Because it is so far from a state of equilibrium it greedily absorbs everything that comes to meet it. This water lacks qualities and what Schauberger calls 'character'; therefore it absorbs gaseous substances and removes carbones from the organism, causing harm to the human body.

Rainwater is the purest available water with mostly oxygen content, but unsuitable for drinking; neither is snow water. Among other immature/juvenile kinds of waters are deep-well water or geyser waters, both from deep underground sources. They contain some minerals but few gases. Surface water, like that of dams and reservoirs may originally have been mature, but the more it is exposed to light and heat the more it deteriorates.

Groundwater originally has dissolved carbones and trace salts; it is water of a higher quality. We are approaching thus to the best and healthiest kind of water: spring water. Water flowing from healthy springs is only to be found in healthy high-quality forests. Here we find a reality that often contradicts deeply held scientific beliefs, and that Schauberger's discoveries render understandable. The artesian hypothesis of springs so entrenched in conventional hydrology—that springs cannot emerge unless there is surrounding water at higher heights—doesn't hold true for springs which lie far above any significant accumulation of water. A spring on the High Priel, west of Salzburg, gushes forth 300' below the summit located at 6,500'. Another example, among many, is that of a spring on Mount Rareu (Bukowina, in Czech Republic) coming out of a rocky outcrop toward the summit.[94] Schauberger indicated a number of differences between true springs and artesian springs/seepage springs. Artesian springs come from water that enters the ground, hits an impervious layer and drains out under the effect of gravity. A true spring defies gravity. The temperature of artesian springs is also different from that of true springs, generally approaching that of the ground layer and in the order of +6 to +9⁰C, instead of +4⁰C for a true spring. Seepage springs are also generally poorer in dissolved salts and trace elements.[95]

Quality in spring waters, indicates Schauberger, is often shown by 'shimmering, vibrant, bluish color.' Spring water is often deficient in oxygen, but rich in carbonic acid. It should not be drunk directly from the spring because it will seek to acquire oxygen; it is better to drink within ten yards from the source.

The important quality-related aspect, which says much about springs if we want to understand their origin, is the fact they always emerge at temperatures very close to +4°C, and the paradox that spring water can actually flow strongest during the daytime and in Summer. Likewise they also deliver cooler waters and rise higher in Summer than in Winter, happily contradicting purely physical laws.[96]

Spring water, emerging at +4°C, can carry most solids in suspension and gases in solution. This is why in newly bottled spring water-bubbles form on the sides of the glass due to its CO_2. In springs the higher the water rises, the heavier the minerals it precipitates, thus becoming denser.[97]

The abundance of springs has been greatly affected by the treatment of forests in the last centuries. The disappearance of springs is a massive ecological change that has passed under the radar,

---

[94] Viktor Schauberger, Callum Coats, editor, *The Water Wizard*, 90.

[95] Callum Coats, *Living Energies*, 129.

[96] Viktor Schauberger, Callum Coats, editor, *The Water Wizard*, 77, 99. An example quoted by Schauberger is that of a 'wandering spring' in the mountains of Montenegro; in Summer it emerges up the mountain, in the Winter down in the valley. Viktor Schauberger, Callum Coats, editor, *Nature as Teacher*, 67.

[97] Schauberger devised a laboratory experiment with communicating vases to show how water rises in springs, defying gravity (Viktor Schauberger, Callum Coats, editor, *The Water Wizard*, 50–52).

but it doesn't escape the notice of holistic scientists even almost two centuries ago. Speaking about Lake Valencia in Venezuela Alexander von Humboldt (1769–1859) said: 'the clearing of planes and the cultivation of indigo over a half a century has affected the amount of water flowing in, as well as the evaporation from the soil; springs dry up, or merely trickle. Riverbeds remain dry and are then turned into torrents whenever it rains heavily on the hills. By felling trees that cover the mountains men everywhere have ensured at the same time two future calamities, lack of fuel and scarcity of water.'[98] A century ahead of the science of ecology Humboldt was able to articulate some of the finer observations to which we will return with more background information. Adding my voice to those of scientists, I am bewildered that, although I live in rural, forested Vermont, springs are a very rare occurrence, and I don't know of any in my neighbourhood, except for those under a man-made pond. A constant hiker I have not come across any of those; at most a seepage spring which dries up in the Summer season.

The wonder of the formation of springs has not been approached yet because it rests on the understanding of the temperature gradient, to which we will return now.

Water and Temperature: The Anomaly Layer and Temperature Gradients

'Nature's creativity, however, thrives on measured coolness.'[99] Water is the liquid which has the greatest capacity to store heat; it absorbs it and releases it slowly. It also behaves abnormally in relation to temperature; it contracts to a maximum at +4°C (39 °F), then expands again upon freezing. At this anomaly point turbulence is also at a minimum, whereas it accelerates the more we move away from +4°C. At the anomaly point, water attains its highest density, which also means its highest energy/life-force content. The layer at +4°C is so important that is has been called the 'boundary layer', the 'anomaly layer' and with good reason the 'temperature-less layer'. It is no wonder that the ocean waters are most productive at these temperatures, as is the case off the coasts of Peru and Chile, thanks to the cold Humboldt Current.

---

[98] https://www.youtube.com/watch?v=pgvX0QdYI6M
[99] Alick Bartholomew, *Hidden Nature*, 34.

Healthy water will naturally seek to flow in darkness or shade, avoiding heat and direct sunlight. A course of water left to itself will shade its waters with trees upstream, and protect its waters through the shading effect of sediments in suspension (turbidity), which has a cooling effect, when it broadens downstream.

The number one enemy of water is excess heat or over-exposure to the Sun's rays. The behaviour of aquatic plants in waters protected from light and heat points to an important reality about etheric forces. Schauberger observed that at temperatures close to +4°C moss and other aquatic plants' shoots stand at right angles with the direction of the current. They point downstream, as we would expect, when the temperature deviates strongly from +4°C.[100] In other words aquatic plants align themselves not with the physical stream alone but with the etheric stream, which is at a maximum due to spring-water temperature and chemical composition.

A series of other temperature thresholds are worth noticing. At 37.5°C (99.5°F) water's specific heat—speed at which it absorbs or releases heat—is lowest, giving it the capacity to exchange large amounts of heat faster. Once the water temperature rises above +9°C (48°F) the oxygen becomes more and more active and aggressive, increasingly promoting decomposition and encouraging pathogenic bacteria as it rises. Below a temperature of +9°C oxygen is used for growth. At +4°C the carbones in gas form, enlivened with buoyancy, are finely dispersed in the water. The oxygen, by comparison, is very condensed, and it sinks; it becomes inactive. This theme of inverse relationship between the activities of carbones and oxygen is one that will return in many natural observations and in Schauberger's applied technology.

One important notion for all of the activity of Nature and the circulation of water is that of the temperature gradient. A positive temperature gradient approaches the anomaly point of +4°C from above or from below; we will consider from now on the one from above. A negative temperature gradient moves away from the anomaly point. We can say that water moving toward the anomaly point breathes in—acquires more gases, condenses—water moving away from it breathes out. The positive temperature gradient is accompanied with reduction processes, the negative temperature gradient with oxidation processes.

---

[100] Viktor Schauberger, Callum Coats, editor, *The Water Wizard*, 144.

Positive gradients are used by Nature in creating life-forms. With a positive gradient ionized substances are drawn together for productive interchange, oxygen becomes passive and bound by the carbones. With negative temperature gradients the warmed-up oxygen becomes more and more aggressive and attacks healthy structures. Both gradients work together, but for health to prevail in water circulation the positive gradient must prevail, as we will see in much of what follows.

The temperature gradient influences nutrient uptake. How this works out can be seen in the soil and in the tree sap. In the soil, under a positive temperature gradient, usually under the shade of trees, all the various nutrients and salts are deposited well below the ground surface as the water cools to +4°C. In the case of a negative temperature gradient, however, due to heat evaporation and little penetration, the lowest quality nutrients are precipitated at the surface, which not only has dire consequences for soil fertility, but also for the proper formation of trees.

In the tree under a positive temperature gradient the highest quality nutrients are precipitated last as the sap cools towards +4°C or is maintained at this temperature. Under a strong negative temperature gradient—as in the case of trees exposed directly to the heat of the Sun—the opposite takes place and only the lowest quality nutrients are expelled; the highest quality ones are not transported at all.

Life emerges at the intersection of contrasting gradients, but for long-term sustainability the positive gradient must predominate. Alick Bartholomew sums it up thus: '… if the positive temperature gradient is very powerful, then the effect of the reciprocally weaker negative temperature gradient is beneficial and promotes the outbreath into physical form of the highest quality substances'.[101] In the contrary instance the outcome leads to substances of inferior quality. In agriculture, in forestry, in river management, dams and reservoirs, engines, industrial and domestic heating, the pattern has been the same, replacing positive gradients with negative ones, or further enhancing the latter.

Once more Schauberger was able to observe amazing phenomena as one approaches the anomaly point. When two almost identical pebbles rub together under cool waters, a fiery glow is clearly perceptible. In this phenomenon, called 'triboluminescence', the

---

[101] Alick Bartholomew, *Hidden Nature*, 117.

fire is not extinguished by water. This energy doesn't cause any electrical disturbances in electronic devices, indicating it's another kind of electricity than the one we know, a bioelectricity we could say. Thus, a kind of energy is freed, but one of which science has little understanding, because it cannot apprehend polarities. Our explorations will approach this other type of energy in deeper and deeper ways.

Planetary Motion

We can best understand the natural motion of water if we refer it to the movement of the Earth, which it mirrors. The Earth spins on its own axis, engendering day and night (rotational motion) and around the Sun (orbital motion) to give us the seasons; the solar system moves within the galaxy giving rise to astrological ages. The liquids on the Earth's surface mirror these movements. This is what is called in mathematical terms a 'cycloid-space-curve'. This is why Schauberger refers to the natural movement of water as 'planetary motion'.

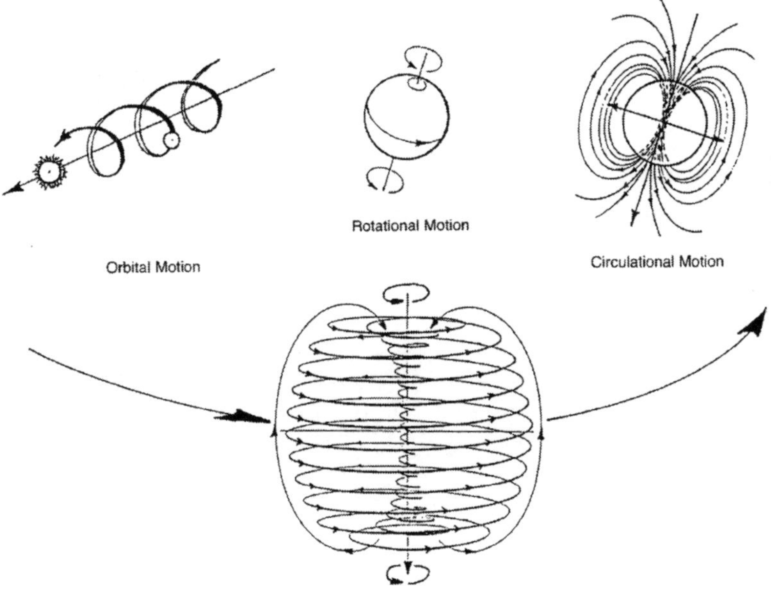

Figure 8: Three basic forms of motion and their combined result

Water in a riverbed mirrors these movements:

- It rotates on its own axis: a motion, similar to the axis of rotation of the Earth.
- It meanders as the Earth spirals around the Sun.
- The whole spirals in a space curve, mirroring the whole movement of the galaxy.

Combining the movements, the river's core-waters will flow in longitudinal vortexes which have a cooling effect when the river is healthy. This movement will alternate between clockwise or anticlockwise as the river meanders from right to left and back to right. Another example: in a vortex produced in a glass column of water in laboratory conditions, water naturally swirls around its centre anticlockwise in the northern hemisphere. Moreover, it will display oscillating rhythm of movement of the central vortex itself and finally a breathing rhythm of the whole. Altogether it gives rise to the 'cycloid-spiral space-curve', or more simply from now on, 'planetary motion'. In a flowform you can observe how the water oscillates rhythmically in each basin, but also how a rhythm and periodicity is established in the movements between basins.

Vortex and spiral are for Schauberger the keys for understanding all creative movement. The vortex has a clearing, purifying and energizing effect on water. It brings it back to a nascent state where it gets rid of all previous negative influences. It begins the process of creating greater diversity and complexity. This principle is used in homeopathy where substance is brought back to its nascent stage, which expresses itself as formative force.

When we look at a river we can also recognize two broad types of water flow, which form a polarity:

- Laminar: the stratified and unimpeded flow of water down an inclined plane. In this kind of flow the layers of water run parallel to each other without mixing. It is said that this is most often the case with low speeds and in fluids of high viscosity. But in water laminar flow is strongest close to the boundary layer of +4°C, even at high speeds, as we have seen. In this instance water does not accelerate.
- Turbulent: at present this is attributed to mechanical effects alone, e.g., in the mixing of different kinds of water. But turbulence also increases with temperature and negative temperature gradients.

Schauberger adds that 'turbulent phenomena in water are nothing less than the counter-motion to laminar flow—arising from physical causes and generating vertical currents in flowing water, maintaining the steadiness of the descending flow through the creation of transverse currents'.[102] Turbulence has a braking effect on the flow of water. In what follows we will see how the two types of flow interact in the longitudinal and cross sections of a river.

Two maxims are central to everything that happens in water and by extension in Nature: 'All life springs from water' and 'All life arises from motion.' And since there would be no life without death/decay we can expect to see two completely different kinds of motion, as there are in fact. One produces the conditions for life and another creates the preconditions for this to happen. One of them, planetary motion, produces the physical forms, whereas the other initiates the processes through which the products of the first form of motion can be brought into being. The second one, producing decomposition is the one we are more familiar with, a motion akin to analysis; the first kind of motion is one of synthesis. The two together form a 'biodynamic movement'. Schauberger qualifies the motion of synthesis as the one that overcomes gravity and brings about the 'densation' of matter.[103] By densation he means the refining and energizing of matter, which leads to the release of levitational/etheric forces.

In what we just explored Schauberger hits upon the differentiation between the forces of levity (etheric forces) and those of gravity. In geometrical terms the difference has been made clear by George Adams with the contrast of space and counter-space. In an example of this we can characterize a line as the sum of an infinity of points in both directions. This is the current view of space.

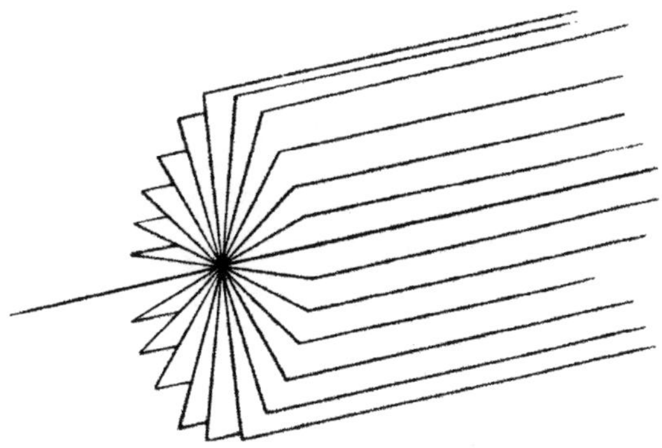

Figure 9: The line as the meeting place of planes

---

[102] Viktor Schauberger, Callum Coats, editor, *The Water Wizard*, 138–39.
[103] Viktor Schauberger, Callum Coats, editor, *The Energy Evolution*, 127–28.

From the counter-space perspective, a line can be seen as the encounter of planes coming from the periphery and intersecting / converging towards the infinity within, the line itself. If we take the definition of the point from this perspective we can say that this is the intersection of infinity of lines originating from the periphery and converging towards an infinity within.

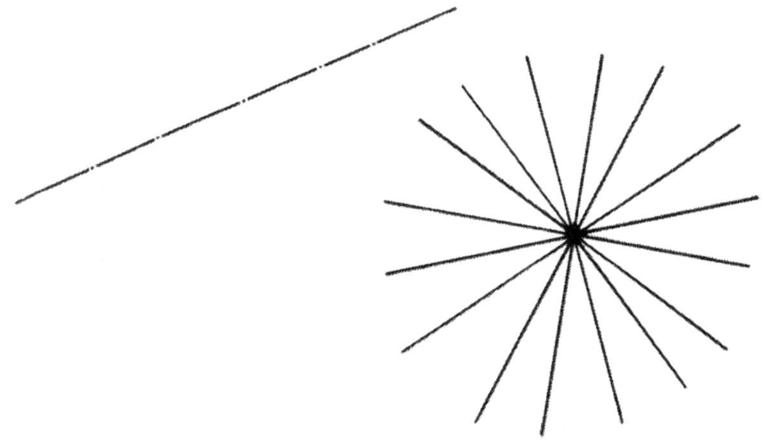

Figure 10: Points to line and lines to point

One perspective gives us the origin in the centre and the infinity without, the other the origin without and the infinity within (in the point). The example of the vortex corresponds to the second kind of perspective; the eye of the vortex or of a tornado is an infinity within, the point in which all of energy concentrates from without. Viktor Schauberger made these ideas perceptible in the motion of water and in the generation of energy.

In essence Nature uses two kinds of motion: the one with which we are completely familiar, which is centrifugal (having an infinity without). The one that only Goethean/phenomeno-logical, or spiritual science, can apprehend is centripetal (infinity within). To the first we will add the motion that is human-generated in technology. It is a force of decomposition but modified and accelerated by recourse to very high temperatures. This is why Schauberger also refers to centrifugal motion as 'technolog-ical motion' and centripetal motion as 'planetary motion'. Each kind of motion generates a set of resulting conditions. Let's look at these in more detail.

The centrifugal motion used in technology generates heat and decomposes. It fragments, creates chaos, noise and heat, in essence it increases disorder/entropy. Almost all of our technology is based on explosion and generates heat and entropy. What happens in controlled manner in Nature at moderate temperatures is accentuated in technology at far higher temperatures.

With centrifugal motion and the prevalence of oxidation one obtains over-acidification, leading to products of emulsion like fatty acids bound by oxygen. At temperatures above +40°C a spark will cause immediate combustion. Pressure turbines are good examples of such motion, where degrading, electrolytic forces develop, causing effects similar to what happens with electricity through air or water. This motion, if acting alone, would lead to sterility and inability to reproduce.

Centripetal, or planetary, motion is a vortical form of motion, moving from the outside to the inside with increasing velocity, which acts to cool, to condense, to structure, assisting the emergence of higher quality and more complex systems. This motion is inwinding, as what we have seen in the vortex; it 'narrows towards a point', the infinity within of our theoretical considerations. It is an upbuilding kind of motion, which is silent and cools off substances. In Nature this can be seen in the in-winding movement of the cyclone, whose centre is cool. Planetary motion densifies structure; it is creative, convergent and formative. In this kind of movement suction increases and friction is reduced: speed and energy increase automatically.

Without levitational forces there would be no circulation of sap, blood or water. These are not raised mechanically, but 'sucked up' by levitational forces, which move according to planetary motion. The densification is not a phenomenon limited to physical space alone. On the contrary, matter is raised to a spaceless condition; it is etherealized. Schauberger explains: 'The relation between the material, energetic and more subtle worlds should be perceived as a pyramid, wherein coarser, less energetic matter occupies the lower portion. As the volume reduces with height [rising in the atmosphere], the proportion between matter and energy gradually reverses until at the very apex all that is left is extremely fine matter or energies in a subtle or etheric state ...'.[104]

---

[104] Viktor Schauberger, Callum Coats, editor, *The Fertile Earth*, 22–23.

In centrifugal/technical motion the resistance to movement (friction) increases by the square of the velocity whereas there is no resistance to motion in planetary motion. Here, Schauberger claims the resistance falls 'by the square of the velocity of a falling heat gradient', defying the law of conservation of energy.[105] To render this understandable we must remember that planetary motion has a cooling effect, bringing water toward the anomaly point. The lack of friction is due to the strengthening of the longitudinal vortical movement in the core-body of water and the release of substances from the walls of the pipes, or banks of the river, as we will see shortly.[106] In Schauberger's work this was demonstrated in the 1952 experiments on water pipes conducted with Professor Pöpel.

For the purpose of all that will follow it is important to understand the ultimate implications of the two kinds of motion. Planetary motion enhances quality and the production of etheric forces. It moves towards what Hauschka calls dematerialization. It does in effect transform substance in the same direction of homeopathy. This is no wonder since homeopathy itself transforms substance through the vortexing movements of potentization—an example of planetary motion.

In natural systems planetary motion also favours those species that depend on high quality environments (e.g., trout), or plants that depend on pure water (e.g., watercress), rather than carps, bottom feeders or algae. Even at present, we can see that where water is not polluted, the riverbanks are shaded and/or the river has not been regulated, the trout survives; where these conditions have changed, as in the vast majority of streams, it will no longer be found. The contrast can even be seen between the upper basin and lower basins of a same river, when conditions change. Centrifugal motion—as we will see in the example of a poorly regulated river—will eliminate the more demanding species and encourage not just algal infestation, but also parasites. Lower, more ancient and less demanding, forms of life will predominate at the expense of those that are present in the best-preserved ecosystems. It was

---

[105] Viktor Schauberger, Callum Coats, editor, *The Energy Evolution*, 74. Here Schauberger is probably referring to the phenomenon of cooling which takes place within a mass of water or air, when submitted to planetary motion. The faster the substance cools off, the more friction is overcome.

[106] Viktor Schauberger, Callum Coats, editor, *The Energy Evolution*, 74.

not hyperbole that brought Schauberger to equate the fate of the trout to that of our culture in general.

We are finding ourselves at a pivotal time of societal choices. Schauberger, like most representatives of Goethean and/or spiritual-scientific natural sciences, underlines the importance of qualitative factors in Nature. He will show us how the movement of water in a river depends on delicate balances of temperature gradients and upbuilding motion in order to preserve higher quality.

The conversation and choices around qualitative factors in Nature affect the very own fate of these factors in ways that can only be perceived … when we honour the qualitative element. In other words, a scientific perspective of input/output, conservation of energy and matter, inexorability of the downward pull of entropy, denial of the etheric forces, will carry a downward trend in which quality will decrease, but such a decrease will not be perceived because … it is not considered important in the first place. Much of the matter of river conservation, which has gone undisturbed and unquestioned for centuries, confirms this trend. All quality is almost absent from major rivers of the northern hemisphere because of the prevailing assumptions of modern hydraulics, which does not recognize the effect of subtle influences and does not address the question of quality other than in marginal ways.

River Health

A healthy, naturally-flowing river creates the best conditions for its banks and the vegetation it needs to enhance its vitality and stabilize its course. When that is the case it maintains its energies within bounds and rarely overflows.

A river follows on its path to lower ground a series of roughly three stages:

- *Youthful Stage: immature water*

  Immature, cold water is hungry for minerals. It gets actively oxygenated in rapids and waterfalls and takes on generative energies.

- *Intermediate Stage: mature water*

  The flow slows, some of the heaviest matter is deposited. A natural river at this stage is protected from excessive warming by the trees on its banks and it recharges the groundwater.

- *Final Stage: the plains*

  The river starts to create meanders and floods, leaving behind oxbow-crescent ponds. The soil is re-mineralized and becomes much more productive. Differences of temperature become very subtle, changing the water gradient, and influencing whether the water removes, transports or deposits sediment. Here turbidity plays an important role in protecting the deeper strata from direct heat, so that they can retain carrying capacity and prevent flooding.

Over the course of its flow to the sea there is a natural pulsation in segments of rivers between sections with positive and negative temperature gradients. In a positive temperature gradient stretch the water gradually heats up after a while, giving way to a negative temperature gradient section. When that happens it deposits sediments and forms barrel vortexes immediately downstream, with cooling effect which allow the stream to pick up sediments anew. This amounts to a natural pulsation, a breathing with the positive temperature gradient as the inbreath, with matter being absorbed, and an outbreath in the negative temperature gradient part. A more holistic regulation of the river amounts to increasing the length of the positive gradient sections. When that is done danger of flooding is reduced.

Another important factor plays a role in preserving the natural pulsation of a river; this lies in understanding the forming of its bends. Here we see a superposition of various vortical movements, which create polarities, primarily three among them:

- Longitudinal vortex generated at the river-bend. The coldest water closest to the centre moves fastest pulling the outer water layers. This is called the 'core-water' where the most vitalizing elements and energies accumulate in emulsion. This is where cold fermentation occurs in which the oxygen, become more passive, combines with carbones which produce a growth-promoting effect. The outer water's turbulence clears the core-water of its silt, and receives from this trace elements and nutrients. Through longitudinal vortexing matter is brought through 'extreme densation' into a physical condition of emulsion rendered possible by high states of ionization, enhancing the generation of levitational energies. The energies are then released on a plane perpendicular to the axis of the vortex.
- Transverse vortexes, which operate at right angles to the banks, are generated by layers of water moving at different speeds. These mix

the water, and at the same time cool it, distributing the lighter weight sediment from the centre to the riverbank. They act as a break, slowing the river down. The uppermost vortex-train manifests as the backward-breaking ripples seen on the river's surface.

- Vertical vortexes directed toward the riverbed. These act in destructive ways in energetic terms. Here oxygen is heated. (See Figure 11.)

We can look more closely at how the longitudinal vortex takes place in a healthy river's core-waters. Let us envision the scenario of a shaded side and a sunny side of the river. The factor influencing the formation of bends is the difference of temperature due to the fact that on one side the water is more shaded and therefore cooler. When it is exposed to the Sun (right side from top to bottom) it becomes more turbulent and decelerates compared to the core-water. The water flowing on the opposite, shaded bank, which is cooler and faster overtakes the slower moving water and curls around it toward the right, eventually creating a bend. The colder water removes sediment on the side it approaches, while on the other side sediment is deposited. Where the colder water flows the channel grows deeper.

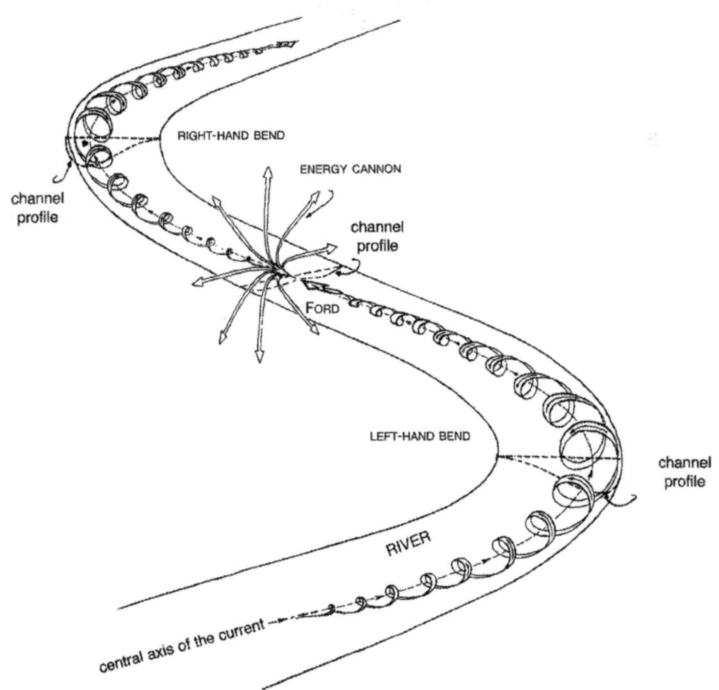

Figure 11: Currents and generation of energy in riverbeds

The cold water will end up flowing on the opposite side of the channel (shaded side) leading to the formation of a bend in the opposite direction to the first, the whole forming the familiar pulsating rhythm. At the bends rocks and stones are ground down, delivering what Schauberger called the 'river bread' with vital nutrients moving into the groundwater table or into the main longitudinal vortical flow, as long as there is a positive temperature gradient. If the water is cold there will be release of ionizing energy which leads to phenomena of triboluminescence produced by the rubbing of crystalline stones of similar composition.

Through longitudinal vortexing matter becomes denser and levitational energies are produced. The energies dissipated when water cools off and the oxygen becomes passive, are released on a plane perpendicular to the axis of the vortex.

Where the core-water starts oscillating from one side to the other the speed of flow tends to decrease; this is where the river is shallower and fords are formed. In the bends rocks and stones are ground up and carried along. At the ford not only solid matter but also upbuilding, levitational energy is released into the environment; this is what Schauberger calls the 'water cannon'.

Overall the water in the river first moves into a bend with increasing energy, then it exits from it to come to a standstill, where the energy cannon arises; on the other side it exits in the contrary direction to start the process again in mirror fashion. All of this resembles what we know of the stirring of a volume of water containing horn manure or horn silica (BD 500 or 501), first in one direction, then in another. Nature's alchemy is replicated at a small scale through the processes devised by Rudolf Steiner.

Technological versus Holistic Water Regulation

River regulation purports to improve the use of flowing water for navigation, for protection of riverbanks against flooding and rupture, use of drinking water or production of energy through reservoirs and hydroelectric power. Basic modern river regulation practices aim at bringing water into its fastest flow through the landscape and at letting it drain as quickly as possible. To do this it modifies/straightens the course of the river and stabilizes it with artificial structures, through canalization and bank-correction, or eventually dredging its sediments. Add to this dams and reservoirs. Basically these measures correct Nature without understanding it. By correcting the consequences rather than the causes they lead to compounded problems. The waterways are much better and more cheaply regulated through the temperature gradient, its own engine of motion and driver, than through the bends, which are only the consequence of this motion.

By the time Schauberger pointed to these problems, practically all major rivers in Europe and North America had undergone a technologically-approached change of their course. The control of the Danube, as of 1931, engulfed considerable sums of money. The waterworks in the upper and middle parts of its course had already rendered unproductive almost a million hectares of farm-land, periodically flooded. In Serbia whole villages had to be evacuated.[107] Schauberger, deeply involved in the consultations, even if on the margins, devoted time and energy to the two main European arteries, the Rhine and Danube, but to no avail. The following is an example among the many from the Rhine. Upstream of Mannheim the meanders of the Rhine were straightened and made to flow in a V-shaped channel of uniform cross section. The course of the river was shortened by half over what are presently about 20 kilometres. This means that the river is no longer able to transport its sediment and the bed has to be constantly dredged. Through this and other similar works the Rhine was robbed of its specific character and vitality, because water is prevented from creating differentials of temperature—it is artificially kept at uniform temperature—and no cooling can take place through longitudinal vortexes. Transverse and vertically aligned vortexes will prevail. A negative temperature gradient becomes the almost constant norm, and the sediments are no longer ground down, nor beneficial energies released. The oxygen becomes active and aggressive and pathogenic bacteria thrive. Having nowhere else to go the water becomes more aggressive in times of floods. Multiply this phenomena over most of the course of the Rhine and you will have a fuller impact of modern hydraulics on the matter of river regulation.

The disastrous results of modern watercourse regulation come from only taking into account mechanical factors and neglecting the reality of the patterns of flow in naturally regulated rivers in

---

[107] Viktor Schauberger, Callum Coats, editor, *The Water Wizard*, 170–71.

which the temperature gradient plays a paramount role. Problems in watercourses are exacerbated because water is made to flow in ways that are not natural to it, resulting in negative temperature gradients, acceleration of the flow, erosion of the banks, lowering of the tractive/carrying capacity of the waters which is optimal when close to the boundary layer, etc.

The above is all the more so when this completely technological approach ignores the roles of forests and lakes in the stabilization of rivers. The shading and the cooling effect that accompanies the evaporation created by the presence of forests would facilitate the emergence of a permanent positive temperature gradient. With this the river would flow more naturally and easily remove and transport sediments in the core-waters at its centre.

Lakes, on the other hand, create a delaying effect on rainwater discharge. Where forests and lakes are missing, then properly operated artificial reservoirs are important, even essential. Here lies another aggravating factor: most modern dams create conditions for purely sterile water which does not recharge the groundwater and causes additional problems downstream. Add to these the problems caused by the heating and lower quality water created by electrical turbines, an issue little acknowledged to this day. The waters exiting the turbines lose all their structure, heat up and become rich in oxygen, which becomes very aggressive in the river's path downstream.

Reducing danger of flooding

Flooding is more likely to happen in negative temperature gradient stretches of the river. To prevent it we must recreate a positive gradient or extend its duration. This can be done in at least four ways:

- Replanting of trees on the riverbank with tree species with a high evaporation rate to increase refrigeration. The belt should be of 500 to 1000 metres (1650' to 3300') wide.
- Because of modern river regulation dams are needed in order to prevent the constantly accelerating movement of water due to the straightening of the riverbeds. If warmer water is discharged from the top of the dam this will encourage flooding. If too much colder water is drawn from the bottom of the dam this may cause excessive scouring and transport of heavy sediments causing problems downstream. In addition 'cold pollution' can cause havoc to the fish population. Modern dams suffer like their

river counterparts, from ignoring the effects of negative temperature gradients which encourage hot temperatures at the centre of the wall, and jeopardize their structural soundness through phenomena of cavitation—formation of empty spaces in the solid structure. Schauberger designed simple but ingenious patents for drawing the right temperature water from the reservoir, temperatures that obviously vary according to the time of the year. One of his patents creates a second skin on the inner side of the dam, in whose gap water is drawn in from different horizons in Summer and Winter. In both instances this favours a positive temperature gradient toward the dam wall, protecting it from overheating. He also designed special dam profiles and devises for letting water flow over the surface of the dam in order to reduce the contrasts of temperature and preserve the soundness and longevity of the dam, still an alarmingly great problem with modern dams.[108] This second measure is only temporary: after six months to a year the structure of the dam is completely sealed and it is no longer exposed to structural dangers.

- Installing flow-deflecting guides that direct the water at the bends toward the centre, imparting a spiralling movement, and creating cooling longitudinal vortexes—anticlockwise on left-hand bends, clockwise on right-hand bends. The flow-guides are an elaboration and refinement of the slats Schauberger placed in the log flumes. They consist of a precast concrete flowform. The curved surface is fluted with grooves running parallel to the direction of flow. The wider, upstream side of the triangle is horizontal and flush with the riverbed. It directs the waters in such a way that they curl over centripetally, giving rise to a longitudinal vortex in the centre of the channel. In other words it is possible to recreate artificially the longitudinal spiralling movement that happens naturally in meandering rivers. An added advantage of the flow-guides is to gather the dissolved carbones and direct them toward the centre to

---

[108] To measure the impact of river management on dams, see the list of catastrophes, which would be perfectly avoidable, at https://www.worldatlas.com/articles/the-deadliest-dam-failures-in-history.html   and https://en.wikipedia.org/wiki/Dam_failure#List_of_major_dam_failures. Because rivers are regulated in ways that accelerate their flow and destructiveness, reservoirs are necessary. But because the influences of temperature on the structural longevity of dams are ignored, dams and reservoirs become an added liability.

mix with the oxygen which in healthy streams collects in the central axis. The carbones are energized when moved centripetally. The oxygen is rendered more passive by the negative temperature gradient.

- Implanting energy bodies in the stream, anchoring them to the riverbed. They are egg-shaped bodies of the same density as the water, which create longitudinal vortexes. Whereas flow-guides are used at the bends, the energy bodies are placed in the straighter parts of the watercourse.

Water Mains

In the use of domestic water Schauberger devised a type of main which not only better preserves water quality; it enhances it. Its novelty consists in bringing the planetary motion that water can acquire in log flumes or in newly regulated rivers down into the structure of a wooden water pipe. In the old times wood was commonly used. Ancient Romans used wooden pipes or conduits of natural stone. Wood was still in vogue in Europe before the Industrial Revolution and even in New York.[109] Wooden mains were manufactured by the Australian Wood Pipe Company around 1910 with good results. They were tested by the USDA (Bulletin # 376) which vouched for their durability and better water conduction than cast-iron pipe.[110]

In normal mains the structure of water deteriorates due to friction and turbulent flow, which decompose the dissolved trace elements. Carbonic acid is removed, and this leads to a deterioration of the vitality of the water. To address these issues Schauberger designed a pipe made of wooden staves—in essence like a miniature barrel. Inside the pipes were silver-plated copper guide-vanes similar in principle to those of the flumes or of river-bends (see Figure 12). Provided the pipes be embedded in sand—to provide a breathing membrane—and protected from light and heat, the system will outlast steel pipes. Wood will not deteriorate if it's not exposed to the light. Mind you, this is true of naturally grown wood, resonant wood, not the wood of commercial plantations. More about this will be added in the next chapter.

---

[109] Alick Bartholomew, *Hidden Nature*, 157.

[110] Viktor Schauberger, Callum Coats, editor, *The Energy Evolution*, 93.

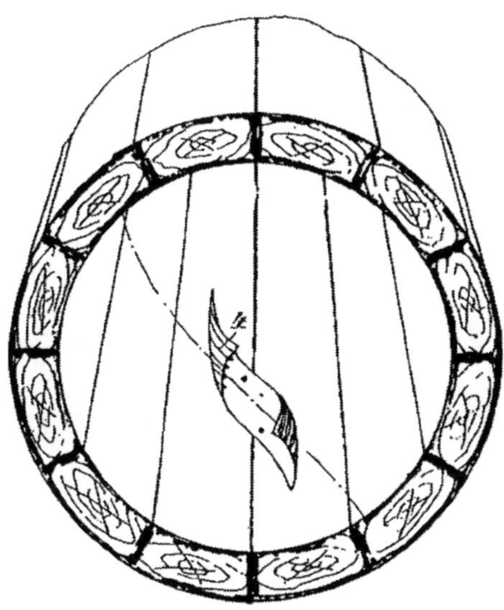

Figure 12: Section of a water main designed by Viktor Schauberger

The guide-vanes in the pipes change the movement of the core-water to a vortical longitudinal one, whereas along the walls a secondary vortical movement occurs that takes place across the section. Hence the name which Schauberger occasionally gave the system, 'double-spiral-flow pipes'.

With this kind of pipe subtle energies are enhanced, friction and deposition on the inner wall are reduced, while the flow is accelerated which has an added cooling effect. The oxygen transferred to the pipe walls helps control pathogenic bacteria which thrive with lack of oxygen. Vortical motion enhances levitational forces and generates tractive force—the capacity to remove particles in suspension (see Figure 13).

The circulation process is dynamic and evolves over time, at least at the beginning. With the cooling occurring in the middle core, some solid particles are directed to the periphery, where they will contribute to seal the wood surface in such a way that the wood becomes even more durable than iron.

As an added bonus the water circulates faster than in regular kinds of pipes and increases in quality, whereas in a regular pipe the water loses quality with the distance of transport. Part of the process of quality build-up is due to the cooling process in the core-water due to gases evolving from the carbones formed in the flow axis, in higher concentration than in the periphery. Oxygen acts in reverse to the carbone gases, concentrating towards the wall of the pipe which is warmer. Over time the processes reach an equilibrium point; the wood cures becoming more and more resistant to external influences, further promoting the water's health.

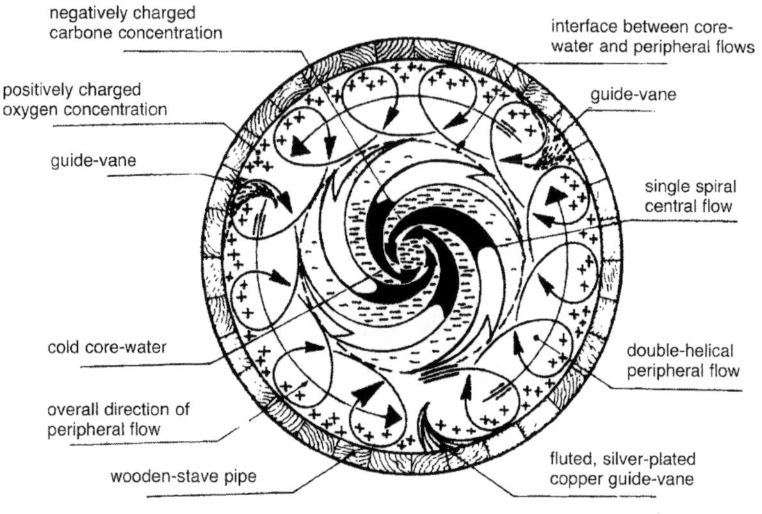

Figure 13: Flow dynamics of the water main designed by Schauberger

Understanding of the importance of the temperature gradient and of the kinds of motion in water has led us to the realization that all of Nature is alchemy, or the movement between energy and matter and back to energy. At the level of the seed, nascent matter shows us how energy transforms matter: witness the experiments of von Herzeele and Hauschka. At the end of the cycle, when the tree raises matter up the trunk, the substances are more and more refined and transmute into energy as we will see next. Homeopathic, or rather biodynamic, processes are everywhere at work in Nature and deliver health to plants and the environment and through these to the human being. If we humans can recover the wonder to which Nature calls us, and think 'an octave higher' we then will realize that planet Earth is not a closed system, and that abundance and health derive from making recourse to other processes than those to which we routinely turn; to move from explosion engines and mechanistic thinking to new technologies and truly living thinking. Armed with the understanding of motion and temperature gradients we can now understand what happens in forestry and farming and how these, together with the rivers, influence the global cycle of water and the climate.

Once we understand even the interplay of temperature gradient and water motion we have access to means of generating practically free energy. Here follows one example of Schauberger's innovations.

Generation of Energy from Seawater

The knowledge of the importance of the temperature gradient and water's chemical composition can be used for energy-generation purposes in the most basic way through solely mechanical effects in exploiting the rising of seawater from the depths of the ocean and its expansion through its recharge with oxygen. Waters from the depths have minimal oxygen content since this has been used up by fish and other life-forms.[111] The water at the $+4^0$C layer, not exposed to light and heat and under conditions of great pressure is unable to absorb gases. For the purpose of generating electricity from the motion of water Schauberger makes recourse to a very long shaft, plunging into the seawater, and connected above water to a specially designed centripetal impeller (rather than a conventional centrifugal impeller), coupled to an electric generator. At an empirically determined depth—that of the $+4°$C boundary layer— the water in the shaft receives, through side conduits fitted with a one-way diffusing filter, an inflow of oxygen (see Figure 14). The oxygen absorption will cause both warming and expansion. The pipes will be designed in such a way that through vortex-inducing vanes, similar in concept to those the inventor used in the log flumes, the water will expand and rise in a planetary motion.

The rising water moves the impeller and sets the generator in motion for the production of energy. The generation of energy can be afforded for truly negligible investments when compared to conventional generation of hydroelectric energy. The end product, oxygenated seawater is harmless to the environment. Of all the energy-generation mechanisms devised by Schauberger this is the simplest, but it already announces a theme: if water or air can be treated according to natural motion and with an understanding of the temperature gradient, then a kind of energy can

---

[111] The whole process is explained in more detail by Callum Coats (Callum Coats, *Living Energies*, 136–38).

be created in ways that do not generate entropy and whose end products are ennobled forms of the base materials, rather than degenerated, polluting forms. This is a 'virtuous cycle' rather than the vicious cycle of fighting Nature with combustion processes which do not occur naturally other than in breakthrough mechanisms, but in much milder forms.

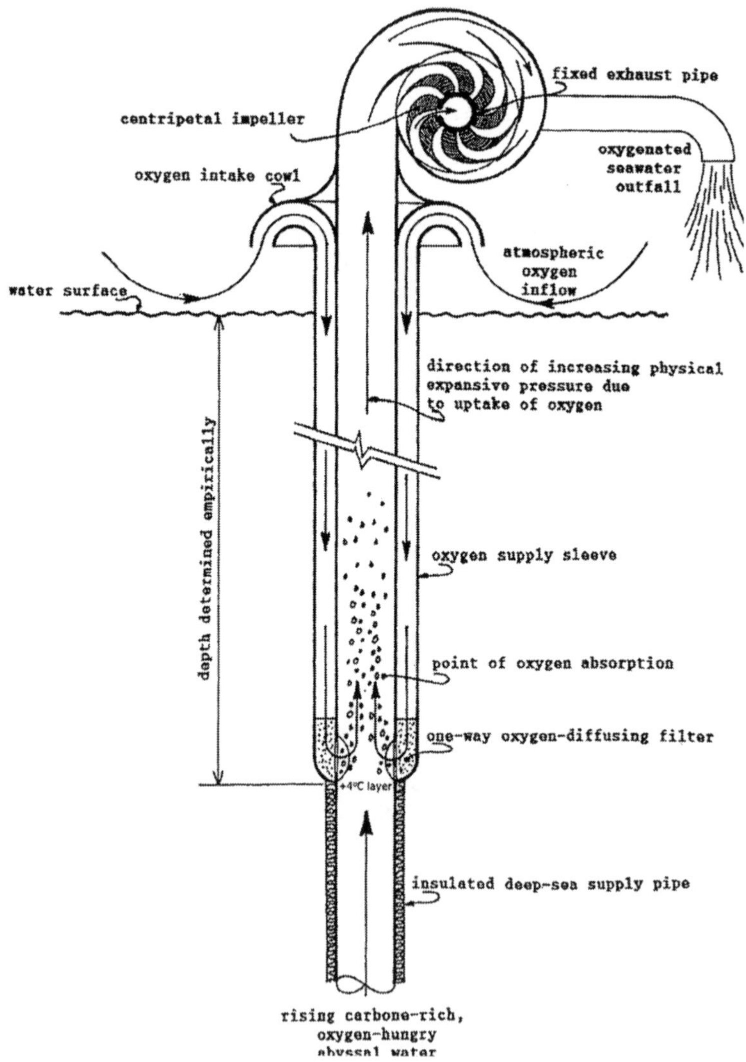

Figure 14: Generating power from seawater

With knowledge of the ideas we have explored Schauberger offered remediation for waterways, simple ways to improve water quality and avoid all the problems that modern hydraulics hardly knows how to approach. The reason behind this is the lack of appropriate understanding of temperature gradients and motion.

Chapter 4

Rivers, Forests, Farms and the Cycle of Water

'The natural habitat of water is the forest. If water's habitat is destroyed, it becomes unstable and hunts about everywhere, ceaselessly seeking out those substances needed for the growth of vegetation.'

Viktor Schauberger

We can either understand Nature as a set of compartmentalized physical units, or see larger dynamics of complementarity and interdependence. We can either sum the parts to build a larger mechanism, or understand the polarities to see states of equilibrium or imbalance of an infinitely complex, but not incomprehensible whole. Water and forest are intimately linked in many more ways than one, making it impossible to create separate compartments and overall models. And most of these relationships are qualitative and non-linear. Our larger view needs to build on an already complex understanding of many aspects of water, as was done in the previous chapter. The whole can then be apprehended in its interconnectedness and, most of all, we can understand what will bring it closer to, or further from, a state of balance and health.

Already from the heading on river regulation we can surmise how many of the modern climate problems are compounded by our lack of understanding of the natural flow of rivers and the complement that forests naturally offer to the riverbed, to the replenishing of the water table, to the quality of the water, to the healthy transport of sediments and maintenance of the riverbanks. All of these add up and have a concrete impact on climate. The overall understanding can only come about through the interrelated importance of motion and temperature gradients. Thus we can at least understand what will bring us closer to a state of balance or further away from it.

The naturally diversified forest is the cradle of water. Its destruction, or artificial management, precipitates alterations of climate at the regional level, which were already noticeable in the 1930s in Europe to those who knew what to look for. Schauberger already recorded decades in advance over climate scientists: 'It should be noted that formidable climatic changes will occur if, as a result of incorrect systems of forest management and river regulation, the orderly formation of clouds is disturbed. Where these systems have been implemented, the number of thunderstorms has consistently decreased, while those that do occur are becoming more dangerous.'[112] Here we can notice that climatic changes are expressed in ways that are different from what we hear most often.

A basic misunderstanding of the forces at play in Nature has brought us practices whose effects are largely underestimated. We live in a world in which, by and large, natural forests have been replaced by plantation forests, or secondary regrowth exposed to the Sun after clearcutting. The overall effects are not very different whether the land is left to natural regeneration or is exploited as a monoculture, though they are more drastic in the second instance. The key difference lies in whether the trees grow under the protection of other trees and the presence of positive temperature gradients, or are left unprotected to the Sun and heat and the effect of negative temperature gradients as we will see shortly.

The systemic problems at play in forestry are mirrored in all of modern agriculture. It is no wonder that Schauberger addressed these as well in ways that go beyond organic agriculture in the direction of permaculture and biodynamics. Agricultural problems too can be seen in relation to extent and types of exposure to light and heat, and how geospheric and atmospheric—Earth and Sun— influences interact and what kinds of temperature gradients are established. Once we understand the place of rivers and forests, the role of farming in climate alteration can also be placed in perspective. Add to this that we can then understand what measures can be used to re-establish balance.

Against these embracing views the exploitation of fossil fuels also plays a role, but one that can be placed in a more sober perspective, as will appear in the next chapter. This is why the present analysis started from the central problems of rivers, forests and agriculture—for the most, largely under-emphasized in the present climate change conversation—and will return to the problem of energy generation at the end to cover what is given

---

[112] This was said by Schauberger in 1931; quoted in Viktor Schauberger, Callum Coats, editor, *The Water Wizard*, 142.

key emphasis today through a purely quantitative analysis of climate change. It may serve as a sober reminder that precisely the qualitative approach of Schauberger detected the problem of climate change almost half a century before conventional science. The quantitative approach can only take note of the obvious when the alteration of quality is so far advanced that measuring instruments can record it.

To understand the maturation of water in its course through the layers of the soil we will look at the chemical changes that take place therein and the physiology of plants, chiefly trees, then return to the cycle of water in which trees and forests play a central role.

Movements of Water and Nutrients in the Soil

For the purpose of the growth of the forest it is all important to understand the influence of waters coming from the depths and rising and their meeting with rainwater percolating downward. We will turn first to the mechanisms which bring water toward the surface—to the growth zone of trees—and which allow for the deposition of nutrients for growth.

Water which has penetrated into the depths and arrived at the important thermal zones existing in the interior of the earth, will encounter the carbones in the form of residues of vegetable matter (e.g., coals) present there. Here, due to the temperature a material transformation takes place in the substances that have come in contact with water, a change generating CO_2 and hydrogen. This is summarized by the following chemical equations:

$$C \text{ (carbon matter)} + H_2O = CO + H_2$$

And further the carbon monoxide recombines with water:

$$CO + H_2O = CO_2 + H_2.$$

The freed hydrogen can react with the salts and precipitate metals or liberate acids. The new CO_2 further combines with water causing the release of carbonic acid (H_2CO_3).

$$H_2O + CO_2 = H_2CO_3$$

Through carbonic acid the dissolution of salts takes place, which are suspended and transported at warm temperatures. As the water moves away from the source of heat the salts will migrate upward.

At one point the waters reach a layer where the temperatures induced by the Earth's core on one side and the Sun and rain on the other meet and reach an equilibrium. This is the boundary layer at $+4^0$C. Progressively above this layer, as the temperature increases the lower quality salts will be deposited first and the highest quality matter will be deposited last, but still well below the surface. If the soil is laid bare and the temperature gradient becomes negative, the lowest quality salts and nutrients are deposited close to the surface with the unfavourable outcome of salinization.

When it rains, especially during the growing season, if there is a forest cover, the Earth's interior will be cooler than the water, establishing a positive temperature gradient, which allows the immature atmospheric water to infiltrate with ease. In its path downward it will absorb everything it can, especially the dissolved salts of the upper strata and carry them down, where they will then be precipitated since the underlying water is already oversaturated. The net effect will be to bring down the salts into an area which is still accessible for tree growth.

The depth of the vegetative zone is determined by the position of the anomaly layer. At this point the daily cycle will play an influence, helping the groundwater, and with it the boundary layer, to rise during the day and, with the effect of cooling and densifying, to sink at night. Likewise it will rise in Summer and sink in Winter.

Announcing what will be enlarged upon as we proceed, with clearcutting the temperature gradients are altered toward negative ones and the groundwater recedes. This means nutrients, after an initial phase of surface deposition, are ultimately carried further down, out of reach of the tree roots. The plant will mostly 'feed' on nutrient-impoverished rainwater and have a spongy growth.

The interplay of water from the depths and rainwater leads us to the understanding of a cardinal phenomenon which is present under healthy forests and most of all those that have not been affected by clearcutting and modern forestry techniques. The interrelated health of the groundwater table and the formation of springs, as they have been rendered understandable by Schauberger, is one that should raise anyone's wonder. Springs and their disappearances should be a weathervane for those of us concerned about global climate. It is all the more surprising that the sinking of the groundwater and the disappearance of springs are topics practically completely ignored in the climate change conversation.

The Water Table and the Formation of Springs

The mechanics of water flowing to and from springs offers some puzzling behaviours that do not fit with the views of conventional hydrology. When these are placed side by side with the nature of springs and the quality of their waters, however, a coherent picture emerges. What we have previously pointed out about spring water acquires a fuller meaning under this perspective.

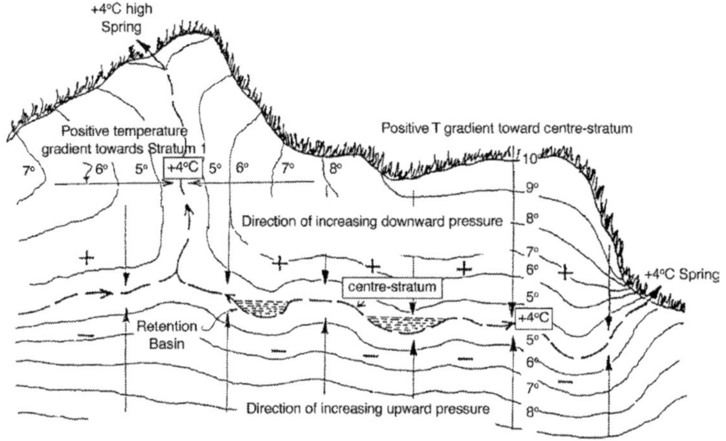

Figure 15: Formation of springs in relation to the temperature gradient

It has been observed that water of a true spring flows at the boundary layer temperature of +4[0]C and does not accelerate when it flows downhill no matter its volume nor the incline of the slope if the riverbed is shaded by trees, in other words protected from light and heat. Likewise, plants growing under such water will not orient themselves according to the current. At a temperature of +4[0]C moss and aquatic plants' shoots stand at right angles with the direction of the current. They will point downstream again the more the temperature deviates from +4[0]C.[113] We can add to this unique behaviour the fact that true springs deliver cooler waters and rise higher during the daytime than at night, and in Summer than in Winter. Here Schauberger's assertion of the two streams present in water finds a confirmation. In waters with strong levitational qualities, such as springs, the physical stream that runs from source to sea is countered by

---

[113] Viktor Schauberger, Callum Coats, editor, *The Water Wizard*, 144.

the strongest possible etheric counter-stream, which goes from sea to source. It is thus that water down a steep slope will not accelerate, and plants will be carried by this stream to the point of countering the purely physical influences of gravity.

Springs arise from the pressure exerted from the percolating rainwater which cools off as it sinks through the ground under a positive temperature gradient. Thus pressure is exerted from above over the $+4^0$C water horizon, while from below another pressure forms from the interior of the Earth forcing the water upwards through geothermal activity (see Figure 15). From the interior of the Earth water rises and cools off, exerting a counter-pressure to the pressure from above. The layer at $+4^0$C, which is the meeting point between the waters from above and those from below, being at its densest, is incompressible and therefore will be forced to move laterally and then, according to the topography, rise vertically. This is why, at least in well-preserved ecosystems such as Schauberger's pristine alpine Austria, it could rise up 300 to 600 feet from the mountain summits, where there is no volume of seepage to speak of, especially not a permanent one. True spring water, being at the boundary layer, can carry most solids in suspension (carbon compounds) and gases in solution, but is deficient in oxygen.

Seen from a holistic perspective two interrelated, energy-connected factors play out in the rising of the spring. On one hand the carbones and female ethericities want to be fertilized/are attracted by the oxygen and male ethericities of the atmosphere. And, as we have seen, at $+4^0$C water is not only at its densest; it is also richest in etheric energy, rising in cycloid-spiral-space-curve motion, thus overcoming the forces of gravity, another factor contributing to the rising of spring water.

The Growth of the Forest

In the following exploration we will refer to the growth of so-called 'shade-demanding trees'. These have a smooth bark and grow up straight. Light demanding trees have a thick, rough bark (e.g., oak) and an outwardly radiating branching habit. In the forests of temperate latitudes we mostly have to do with shade-demanding trees. Their specific needs are not understood by modern forestry, stemming from a complete ignorance of qualitative effects, the simplest one of which is the temperature to which the bark is exposed and the resulting temperature gradient between bark and core of the tree.

The tree lives in the balance of atmospheric and geospheric influences. They breathe through their bark and cannot do so if they are exposed to the light. In a naturally growing forest the atmospheric radiation filters through the canopy of the forest and enters the tree-trunk laterally, continuing its journey toward root and crowns according to the rhythm of the day and of the seasons.

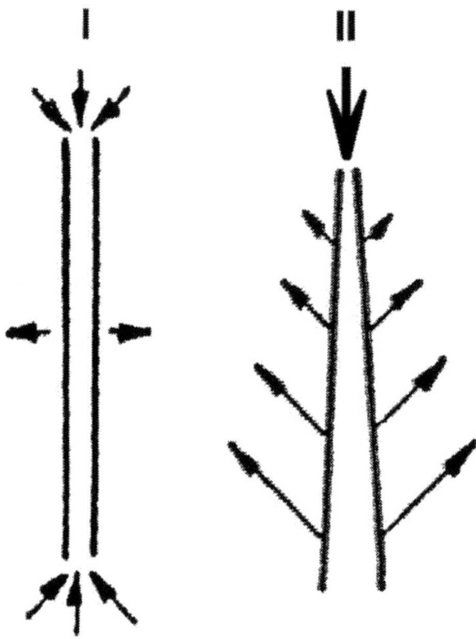

Figure 16: Energy patterns for a shade tree growing in the shade (I) and one exposed to the Sun (II)

What Schauberger observed in his decades as a forester will make more and more sense as we proceed. After speaking of the atmospheric influences he speaks of their geospheric counterparts thus: 'The geospheric rays, on the other hand, enter the tree via the root protoplasm and from there they begin their ascent. These contra-directional, formative substances interact in the space between the base of the trunk and the lowest branches, and it is here that the best quality wood is to be found. Conversely, if vital interactions have been corrupted, especially from above [light-induced growth through clearcutting being the major one] it is here the first symptoms of putrescence appear.'[114] The form of the tree will be

---

[114] Viktor Schauberger, Callum Coats, editor, *The Fertile Earth*, 127.

completely altered. Instead of an almost cylindrical growth with little side branching in a natural forest, we will see a rather conical tree with branches reaching almost to the ground. See this summarized in Figure 16.

In a dense, shaded forest the tallest trees, which Schauberger calls the 'mother trees', or those which grow on the edge of the forest, protect the more delicate and light-sensitive ones which need a CO_2–rich environment and coolness for optimal growth. The mother trees render oxygen a high-grade nutritive material available in the atmosphere of the other trees.

The miracle of life on Earth is continuously renewed by trees. The oxygen which animals and human beings need to breathe is continuously renewed by the forest, whereas the CO_2 is used up, through the process of photosynthesis. During the day photosynthesis converts the nutrients coming up through the sap into carbohydrates and releases oxygen and water. This release further serves to cool off the tree. We see here at play another homeostatic mechanism tending at preserving the positive temperature gradient.

During the day, in the presence of light, photosynthesis can be expressed thus in chemical terms:

$$CO_2 + H_2O = photosynthesis + O_2$$

$$CO_2 + H_2O + light = CH_2O \text{ (carbohydrate: e.g., } C_6H_{12}O_6 \text{ glucose)} + O_2$$

To produce the chlorophyl central to photosynthesis, magnesium, CO_2 and water are required:

$$Mg + CO_2 + H_2O + light = chlorophyl + O_2$$

It is through these processes that, ideally, the proportions of atmospheric gases are kept constant: CO_2 at 0.3%, oxygen at 20.95%, nitrogen at 78.08 %, noble gases (e.g., argon, neon) at 0.93%. Of the oxygen produced by the tree 60% is released during the day, the rest is used up in cool oxidations to produce the substance and structure of the tree itself.

Trees cannot adapt to rapid change; among plant organisms they are the least adaptable. This is why they need the cover of the forest canopy to thrive in growth and health. The movement of the sap is determined most of all by the temperature gradient between bark and core. Under a positive temperature gradient (from outer to inner) in the upward flow of the sap highest quality nutrients are precipitated last and there is a resulting even deposition of growth material with tight annual rings. Growth takes place mostly at the crown.

During the day in the rising of the sap the coarser elements are deposited first into the tree structure. As the capillaries become smaller the sap rises up during the day and goes down at night. As the sap rises or descends in spiral motion, the finer particles deposit toward the leaves or roots. They are accompanied with what Alick Bartholomew calls the 'most highly potentiated homeopathic resonances and amounts of barely structured matter'. At the furthest tips of growth on the tree crown, 'energy radiates into the environment, a process of life giving life, while at the root zone the energetic polarity seems to be that of life seeking life'.[115] Here we are reminded that growth takes place, not because of the deposit of substance, but through the meeting of polar formative forces, such as those coming from the atmosphere and those rising through the Earth.

In a naturally growing forest, during the day there is a positive temperature gradient from the air to the inner sap, which rises and brings up the nutrients and energies required for the most active quality growth. The sap towards the outside of the trunk warms up and expands; it squeezes the cooler sap in the centre, which helps the sap to rise. In the warmed-up sap on the outside of the trunk the male oxygen becomes active, while the carbones (carbon compounds) are more active in the cooler conditions of the inner trunk, where they form starches and proteins.

The reverse happens at night when the temperature gradient reverses. The bark is cooler than the centre of the tree, and the crown cools faster too, causing the sap to sink. The oxygen is released in the centre of the trunk. The coarser nutrients are deposited inside the bark (outermost layer of the cambium ring) while oxygen and gases are moved downwards towards the root zone. The oxygen can then assist in the decay in the humus horizon and stimulate growth at the root tips.

The trunk's structure is divided in xylem and phloem, which play a role in the downward and upward movements of the sap. The phloem's fine capillaries carry oxygen and nitrogen down to the roots; the xylem carries the ionized minerals, salts and trace elements, carbonic acid and CO_2 upwards. This division of the two tissues is found down to the structure of the leaves.

---

[115] Alick Bartholomew, *Hidden Nature*, 197.

In a naturally growing tree, shaded by the forest canopy, in the upward movement of the sap, the capillaries at the top of the tree become extremely small, so that only the finest substances with the highest nutritive qualities remain, the coarser have been left behind to build up the structure of the lower part of the tree. This refinement of the energy at the leaves, with increased homeopathic potency, receives the highly energized drops of falling rainwater. Here takes place an immediate transfer of pure energy or life-force, and therefore growth takes place.

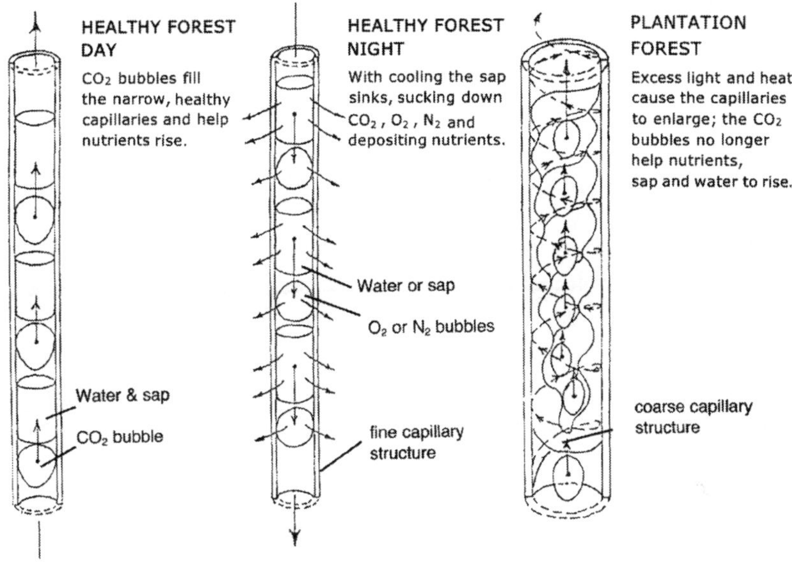

Figure 17: Physiological differences between shade trees grown in the shade (I) or exposed to the Sun (II).

Mechanical suction in the capillaries alone cannot explain how the sap rises at great heights. In a healthy tree the sap is stimulated by pulsating action, the very fine dimension of capillaries and the formation of CO_2 bubbles from H_2CO_3 when the sap is warmed (see Figure 17). The bubbles rise pushing the water and nutrients upward to the crown. At sunset with the temperature cooling the process is reversed as the tree breathes oxygen to bring nutrients to roots and trunk. The bubbles of CO_2 condense and sink. CO_2, O_2, starches, sugars and trace gases formed during the day by photosynthesis are carried down the trunk and consolidate the structure of the tree. While the trunk and crown cool down, the roots warm up, keeping the soil warmer during the night and in Winter, and cooler during the day and in Summer.

In trees that grow in the shade the sap ducts are virtually straight, the growth strongly vertical. Such trees could grow in very deep shade and in poor soil conditions and their annual rings can barely be distinguished by the naked eye. The wood has a 'resonant quality'. It has almost unlimited durability and was thus used for the manufacture of high-quality furniture or the finest, resonant musical instruments, like the Stradivarius violins. This would be hard, or impossible, to replicate with lumber derived from modern forestry practices.

During the day the process that has emerged in the soil and then in the tree, continues in the atmosphere, following similar laws and dynamics. Here too the positive temperature gradient plays an important role. The water that evaporates from the trees ascends into the atmosphere seeking the anomaly layer +4°C, at about 3000–4000 m (about 10000' to 13000'). Here it is once more in a 'taking' mode, ready to take up the finer and more spiritual energies from the Sun and the cosmos.

Things stand in almost reversed terms when a natural forest becomes a plantation. Shade-loving trees of temperate latitudes need to be shielded from too strong a Sun radiation, which must be mediated, e.g., through the establishment of a positive temperature gradient under the canopy of other trees. Otherwise the radiation will reach the tree-trunk directly, and because of overheating the tree will be highly susceptible to parasites and fungi.

In a monoculture tree plantation, under the pressure of the prevailing negative temperature gradients the tree will send out branches even right down to the ground to protect itself, sacrificing upward growth and developing a cone-shaped form (see Figure 16, II). This is the pattern that at present is believed natural in conifers, which are delicate shade-loving trees.

Under commercial plantation conditions, the capillaries are enlarged and the CO_2 bubbles cannot rise with sap and nutrients. This unnatural pattern of growth results in lumber with a lot of knots. Since the sap cannot rise as high as it normally would, there is premature deposition of lower quality nutrients. The highest quality nutrients are not transported at all, causing the crown of the tree to die back, the tree to fall more easily with storms and its lifespan to shorten.

Modern forestry, which does not recognize the growth conditions of shade-demanding trees, fundamentally alters their physiology. It encourages the deposition of nutrient salts of inferior quality in the wrong places. It tends to favour great quantitative output, leading to an unnatural, accelerated growth but of poor quality. What is considered success by corporate forestry—larger annual rings—is actually an aberration for a holistic understanding of tree and forests.

A plantation tree displays a growth turned spongy and coarse, leading to the formation of ring-shakes, in which the wood splits in annular fashion. When cut such wood tends to warp, especially if there is a mix of tight and wide rings.[116] In addition there may be an off-centre heart, heart rot, radial cracks, all of which encourage bacteria and parasites. Such problems were already perceptible with beech lumber in the eastern part of Czechoslovakia in 1931, leading to the formation of a second heart completely different in colour and structure and secondary appearance of white-rot. In Romania similar problems completely degraded the quality of the wood.[117] Schauberger called this compendium of results 'cancerous growth'. He saw illness as a positive factor, as Nature's health police eliminating weak growth.

Modern forestry practices also subvert the soil ecosystem in which there was originally a wide distribution of root horizons according to the species; where soil could be occupied in many layers, whereas now a single layer becomes the norm. The absence of one species affects the others and disrupts the water supply, a problem, which we see at play globally. With monoculture and clear-cutting the water table sinks, the nutrients migrate downwards. The trees therefore develop in drier environments. Since plantations often use conifers, the tree develops a great amount of growth down to the base, which dries off with the subsequent shading of the tree stand. The amount of deadwood of a conifer stand is an obvious liability in case of high temperatures and possible fires. The decreased vitality of the trees manifests in lower growth at the tops and lower resistance to windstorms. Entire stands of conifers

---

[116] During their growth plantation trees are first exposed to the Sun, then shaded by each other, and eventually exposed again through subsequent thinning; thus the growth can be mirrored in initial wide annual rings followed by tight rings and a final return to wide rings.

[117] Viktor Schauberger, Callum Coats, editor, *The Fertile Earth*, 134–35.

toppled are common sights in plantation forestry because weakened trees do not develop healthy crowns and do not anchor into the ground as their healthy counterparts would.

A most promising rebirth of the understanding of the need for the forest comes from permaculture. The use of shelter belts (best in spiral form) lessens the evaporation rate, wind speed and soil's dehydration up to 120 metres (400') downwind and 30 metres (100') upwind, and renders farms more productive according to Canadian research.[118]

What has been said so far about rivers and forests can be followed up further according to the same general guidelines that have emerged in this chapter in relation to agriculture. In farming too modern agriculture has led to the depletion of the levitational forces, the prevalence of negative temperature gradients, the deterioration of water quality and the sinking of the groundwater.

Agriculture and the Depletion of Levitational Forces

What Schauberger saw in relation to rivers and forests continues in relation to the more humanized landscape of agriculture. Here, the Austrian researcher moved in directions that are parallel to permaculture and biodynamics. In fact he used at times the term 'biodynamic', not in relation to the farming approach but in relation to the forces he wanted to support.

In modern agriculture we see tendencies that work in parallel with and reinforce what we have outlined in forestry and river regulation. In modern agriculture too the ground is laid bare and worked with materials and approaches that deplete it of its etheric forces. This was very clear for someone who knew biological growth to be the meeting place of the forces of Sun and Earth, not the result of deliverable nutrients. 'The nurture of the spiritual driving force of life is always more important than the nourishment of the physical form that merely enfolds the soul. The physical form becomes a framework for the spiritual forces of regeneration once it has reached maturity.'[119] And further: 'No plant is actually nourished by dissolved matter, but rather with ascended nutritive

---

[118] Alick Bartholomew, *Hidden Nature*, 186.

[119] See Viktor Schauberger, Callum Coats, editor, *The Fertile Earth*, article 'Increases in Soil Productivity', 157–63 and Viktor Schauberger, Callum Coats, editor, *The Energy Evolution*, 35, 'The Suction Turbine', 32–36.

entities of geospheric provenance in a fourth-dimensional state. These diffuse ethericities can only enter the sap-stream via the root protoplasms, where they are fertilized by diffuse oxygenic ethericities. The higher out-birth of this emulsion (*ur*-procreation) is an ethericity that belongs to the fifth dimension.'

While we haven't looked specifically at what Schauberger calls 'ethericities' we have already addressed the contrast between geospheric and atmospheric influences, which can be seen as the contrast between maternal and paternal energies, or carbones and oxygenes and their role in plant growth at the material level. It is these reciprocal influences and their integration that Schauberger aimed at reinforcing in all his agricultural innovations.

Schauberger, in common with biodynamic farmers, more so than just organic farmers, detected that there was more than negative physiological and product quality problems with the use of artificial fertilizers. On top of these undeniable effects fertilizers cause a whole alteration of the field of energy around the growing crop. Artificial fertilizers, created through the influence of fire, are devoid of potential and rob the groundwater of levitative energies. For these reasons, de-energized artificial fertilizers contribute to the sinking of the groundwater, an effect all in all parallel to what happens to the groundwater below forests under modern forestry management. Schauberger's critique of the disruption of finer energetic influences went also, and for the same reasons, in the direction of the prevalent choices of materials of the Industrial Revolution, which replaced mostly wooden implements with iron and steel tools. He asserted: 'Iron and steel, which have been polarized by fire, are very dangerous to forest and field alike, because these discharged substances attract the valuable soil-energies like a magnet.'[120]

Overall then, Schauberger's contribution to what we can do to reinforce the field of biodynamic influences lies in three general realms:

- Combined choice of form and materials for the preservation of levitational forces in the soil: copper-plated or phosphor-bronze tools and dynamic shapes of these—e.g., ploughs using path-curves designs.
- Generation of levitational forces, making recourse to the conjugation of maternal (Earth) and paternal forces (Sun). These are the general

---

[120] Viktor Schauberger, Callum Coats, editor, *The Fertile Earth*, 124.

principles also used in the generation of high-quality drinking water or etheric energy, that we find at the base of the ideas of 'noble compost' and 'repulsators'.

- Ways of accelerating decay of organic matter whose dissolution is problematic—e.g., liquid effluents from dairy farms.

To explain what brought Schauberger to replace iron and steel with copper-plated or phosphor-bronze tools we must first understand the difference between ferromagnetism and diamagnetism. There are in fact three types of magnetism:

- Ferromagnetism: substances that hold magnetism, corresponding to the natural formation of north and south poles within the substance itself. Among ferromagnetic elements we find iron, cobalt, nickel. At a certain temperature, the Curie point, they switch from ferromagnetic to paramagnetic.
- Paramagnetism: in this case the substances do not hold magnetism per se, but each of their atoms line up in the direction of the external ferromagnetic field. Among these we find oxygen, the air itself, aluminum, osmium and almost all compounds of iron.
- Diamagnetism: detectable in those materials which deflect at right angles in the presence of a magnetic field. Some diamagnetic substances are bismuth, antimony, zinc, tin, lead, water, carbon, copper, silver, gold, glass, carbon di-sulphide and wood. Some of these diamagnetic materials and their positive effects on levitation have already been introduced in the previous chapters. The link between diamagnetism and levitational forces is clarified by Schauberger thus: '[Diamagnetism] comes into being when, under the exclusion of light, heat and air, the media of earth, air and water are made to move radially/axially by means of the cycloid spiral space curve, i. e. screw-form motion from the outside inwards.'[121] The latter expression refers to planetary motion. Diamagnetic materials in effect render possible and reinforce planetary motion.

In our analysis we are particularly interested in the extremes of ferromagnetism and diamagnetism. Modern technology mostly uses ferromagnetic materials; eco-technology makes recourse to diamagnetic material. The first favour gravitational forces, the second the etheric forces.

---

[121] Viktor Schauberger, Callum Coats, editor, *The Fertile Earth*, 75.

The Earth's surface contains 4% iron, which generates a weak ferromagnetic field over the planet and atmosphere. Whereas the atmosphere (paramagnetic) aligns with it, diamagnetic materials move at right angles with it, which means upwards. This explains the behaviour of logs and stones in cold water under the full Moon that we have encountered in Schauberger's observations. It is also diamagnetism that explains why rubbing stones under water generate sparks; these are rocks with a high metallic content. Form and substance contribute to the unusual behaviours, whether it be the rising or the freeing of energy.

In essence the use of copper in farming tools reverses the destructive tendencies of ferromagnetism in iron and steel. In addition we will see in what follows that not only copper, but other diamagnetic materials—zinc, silver, sometimes gold—are used in devises that Schauberger designed for the spreading of levitational influences. When used in farming, these could be considered a parallel and addition to what we know of the homeopathic/alchemical remedies that Steiner offered in the biodynamic approach, among which we count horn manure (preparation 500), horn silica (preparation 501), and the compost preparations (preparations 502 to 507).

Central to the Austrian innovator's approach to farming is the following quote: 'In fact the *ur*-cause of the success of biological methods lies in the right mixture of terrestrial and cosmic substances.'[122] On the basis of this: 'it is possible to produce any kind of water, gaseous or solid matter of a higher order almost without cost, through the appropriate intermixture of animalistic or organic currents'.[123]

The balancing of the two polar opposite forces produces effects in the soil down to the level of the groundwater table, which it prevents from sinking:

> Water … can be charged with gravitating matter from above and levitating matter from below. The logical outcome of this directionally alternating ingress of charging substances is the rising and falling of the groundwater table, whose direction of movement is determined either by the highest-grade transformative products, or cosmic and geospheric rays, originating from above or below. The water will be endowed with a maternal character by being supercharged with

---

[122] Viktor Schauberger, Callum Coats, editor, *The Fertile Earth*, 148.
[123] Viktor Schauberger, Callum Coats, editor, *The Fertile Earth*, 150.

> geospheric, levitating substances. Conversely, the groundwater table
> will sink, if it is overburdened with gravitating substances.[124]

Not only did Schauberger differentiate between ferromagnetic and diamagnetic substances; he also recognized feminine and masculine substances among these: 'Maternally oriented substances are limestone, gold and copper, whereas silicon, zinc and silver are paternally oriented counter substances.'[125] In what follows and in the next chapter we will often see copper on one side and zinc on the other.

The ultimate goal of Schauberger's interventions is to strengthen the etheric influences, which he equates with the formation of a maternal envelope, an amniotic fluid that extends all the way to the groundwater: 'Should humanity desire to increase growth, then it must ensure that the earth's rhythm is strengthened through the artificial creation of material amniotic fluid. … it is therefore possible to infuse the groundwater with unlimited quantities of these growth-enhancing substances. The effect of this is to transform the groundwater into a maternal amniotic fluid,…'[126]

Devises for the Production of Levitational Forces

Let us look at three such ways to create a maternal envelope in the soil: the building of so-called 'noble compost', the use of egg-shaped water cisterns and the treatment of unfermented manure.

Noble composts are egg-shaped, layered composts built around the base of a large fruit tree, benefiting of its shade.[127] This is important in the establishment of a positive temperature gradient toward the centre of the pile, thus favouring reduction processes. No putrefaction will take place. The layers alternate vegetable and leaf waste-matter with others of earth/sand/fine river gravel. Without going into the technical details it is important to know that diamagnetic materials of opposite nature—copper and zinc in the form of filings—are added in the soil mix and that the whole egg-shaped compost is covered with clay or other impermeable material to protect it from juvenile, element hungry, rainwater. The compost

---

[124] Viktor Schauberger, Callum Coats, editor, *The Fertile Earth*, 151.

[125] Viktor Schauberger, Callum Coats, editor, *The Fertile Earth*, 152.

[126] Viktor Schauberger, Callum Coats, editor, *The Fertile Earth*, 152.

[127] For the principles concerning noble compost and how to build it go to Viktor Schauberger, Callum Coats, editor, *The Fertile Earth*, 167–73.

quickly attracts microorganisms during the growing season and is ready for use in the Autumn.

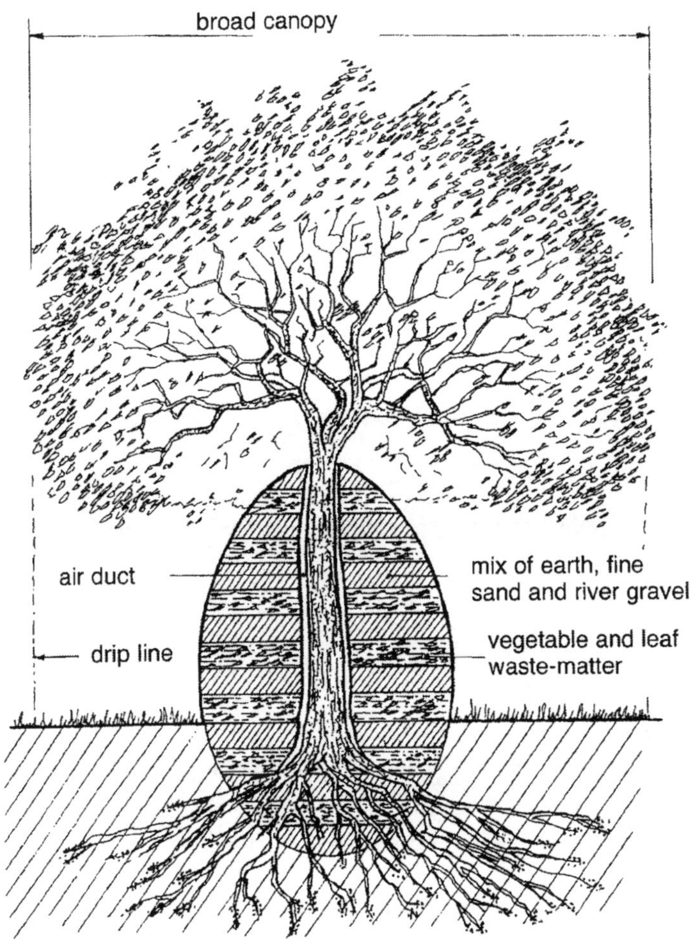

Figure 18: Noble compost

Once matured, the material is spread out as a homeopathic fertilizer, in low doses over the fields toward the evening, counting especially on the help of rain or dew the next day. Layers of ¼ inch or less are sufficient if spread out and ploughed in with the appropriate tools, copper-plated, bronze-phosphor alloy or wooden ones. The material thus produced will have favourable effects on crop yields and on their health.

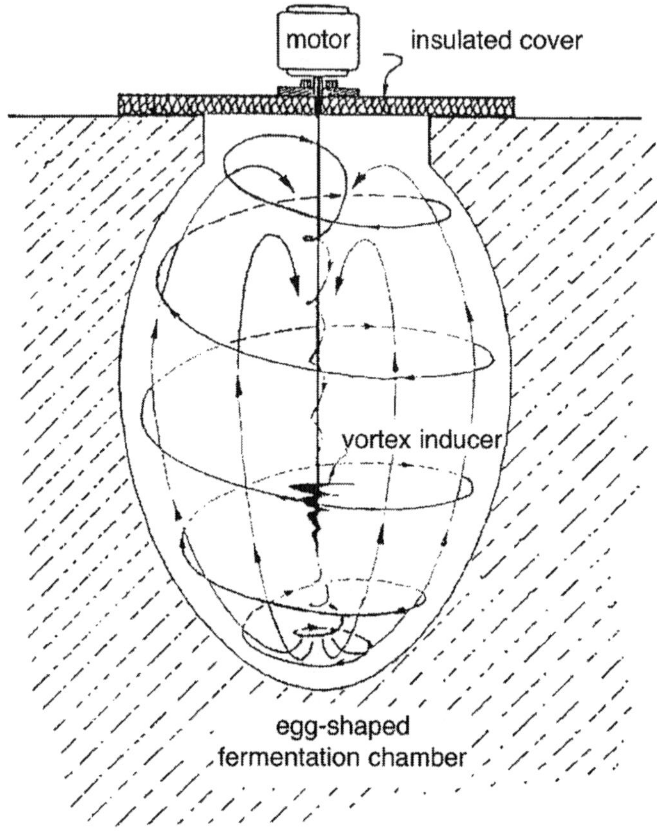

Figure 19: Egg-shaped fermentation chamber

Other devices for the production of levitational forces, or creation of amniotic fluid are the egg-shaped cisterns buried into the ground, insulated and protected from direct sunlight and heat. The simplest ones make recourse to jute sacking to cover the surface opening. They average six-feet deep and make use of organic matter, such as fresh finely broken-down organic waste-matter and liquid manure mixed with oxygenated rainwater.[128] The mixture is left to ripen for a period of about six weeks. Here too we make recourse to copper and zinc filings in addition to the rest. An alternative to this very simple device is one that uses a more permanent egg-shaped cistern (e.g., built of wood) and covered with an insulated cover on whose top lies an engine. Attached to the

---

[128] For the principles concerning Egg-Shaped Cisterns go to Viktor Schauberger, Callum Coats, editor, *The Fertile Earth*, 154–55 and 163–66.

engine is a shaft running through the middle of the cover, which operates a vortex-inducing paddle in the lower third of the cistern. The shaft and paddle are formed of an alloy of bipolar metals—e.g., bronze rod with a silver-plated copper paddle or other vortex-inducing shape. The induced vortex movement accelerates what happens at much lower speed in an egg due to its path-curve, bringing the water to a temperature of $+4^0C$. The resulting product is created in just two to three nights and is similar to the one previously described.

The maternal forces produced propagate horizontally, while the product degrades into an odorless or sweet-smelling product in a way that could remind one of the process at play in the formation of horn manure. Schauberger claims that the resulting fertilizer is about ten times more potent than regular fertilizers. Once more the process has made recourse to cold oxidation and the product, sprayed in the evening is activated the next day when the Sun's paternal energy meets with the maternal energy that has evolved in the cistern. An added benefit Schauberger claims is that with one or two cisterns it is possible to imbue with levitative forces the soil over several square kilometres.

On a larger scale, liquid effluents, such as liquid manure, can be placed in a manure pit, likewise protected from direct light and heat, to undergo cold oxidation through induced vortical movements alternating in one direction and another. In this case Schauberger makes recourse to specially designed whorl-pipes and the process is completed within the order of a few weeks.[129]

We have seen that so many of Schauberger's devices so far make recourse to the effect of form, and all the more so in what will follow concerning energy production (Chapter 5). Can we gain a better understanding of the relationship between form and etheric energy, or life? Lawrence Edwards, whom we have met before, has looked in depth at the question of form, particularly from a mathematical perspective. The relationship between form and substance can be seen in two ways:

- Form can be seen as 'a necessary adjunct of matter': the form is there because of the substance.
- Form can be seen as the primary reality: the substance is there because of the form.

---

[129] Viktor Schauberger, Callum Coats, editor, *The Fertile Earth*, 12–13.

Through what he has documented over decades Edwards concludes 'To me it seems possible that the full reality might take account of both aspects. A bud, at its first appearance, finds itself close to the ideal form; because of this it is able, from the start, to take up strong life forces, which then in their turn stir its progress vigorously further towards the ideal form.'[130] This leads Edwards to see form as the bearer of life-force. And Edwards finds confirmation of this in the work of Rudolf Steiner. When Steiner looks at the first forms in creation, what he finds is the egg. He never calls them ovals, always eggs, leading Edwards to conclude: 'I think we must see it as something which is associated with the very genesis of Time itself.'[131]

All of this brings us closer to the relationship, indeed complementarity of egg and spiral path-curves. Edwards calls the vortex form a 'cup', a term which will lead us to a deeper understanding of the relationship between the forms. Within the realm of path-curves in mathematical / geometric terms the negative of the egg, obtained by transforming the key parameter Λ, is the vortex. We have previously seen that the linear process of collineation—transformation of a volume into itself through three invariant points—generates path-curves through relatively simple geometrical and mathematical operations. The corresponding equations revolve around the key parameter, Λ, which determines the shape of the resulting path-curves (see Figure 5, p. 32). In our instance when $\Lambda = 1$ we have the egg form, when $\Lambda = -0.5$ the vortex. Above 1 the bluntness and pointedness of the egg apices become more accentuated; below 1 forms move toward the vortex. At $\Lambda = 0$ we have a perfect cone. Thus Λ controls the quality of the form.

The bud of the plant, with its enclosed gesture, transforms after its opening into the cup-vortex form of the whorl of leaves or of the flower, and between these two forms Edwards sees a close relationship.

> We have the enclosed, and enclosing, form of the egg, and the bud, enshrining within them the forces of our most elemental beginnings; and we see them, before our very eyes, changing into the form of the cup, the chalice open to whatever the future may bring ... And in geometry, all we need, that this miraculous transformation should

---

[130] Lawrence Edwards, *The Vortex of Life*, 220.

[131] Rudolf Steiner, *Spiritual Hierarchies*, lecture 3 of 13 April, 1909, quoted in Lawrence Edwards, *The Vortex of Life*, 263.

be fulfilled, is that κ should move from the positive to the negative numbers. These two types of form are really one and the same thing. We cannot have the one without the possibility of the other being present.[132]

Edwards goes one step further in linking the two forms in an organic breathing:

> One begins to sense that an increase in κ is associated with inbreathing and tension, coming into a certain climax in the form of the wild rose-bud, whereas a decrease in κ leads to relaxation, and finally, when κ becomes negative, to the bud being transformed into the cup-like form of the vortex, open to the whole universe... The two processes go hand in hand, every big outbreathing being preceded by a little moment of inbreathing.[133]

It is no wonder then that what Edwards has discerned mathematically from a spiritual-scientific perspective is what Schauberger perceived directly. In the egg-shaped cistern the vortex form—occurring naturally in the simplest model or induced in the most sophisticated one—radically transforms substance.

From all of the above we have gained a better understanding of what Schauberger shows we are doing in forestry and farming, and what we should do to reverse the ecological downward trends that we see at present. Much of this can only truly be understood thanks to the knowledge of water's motion in relation to temperature gradients, and of the importance of form in relation to life.

We can now integrate the contents of Chapters 2 to 4 to arrive at the determining factor of climate change, or rather planetary depletion, in keeping with what has emerged from Schauberger's work. When water, forest and river ecology are fully understood then we can come to the crucial recognition of the major source of climatic disruption, the alteration of the cycle of water, which Schauberger recognized as early as the 1930s.

Two Hydrological Cycles

Looking at the ecology of rivers, forests and agriculture is a way to encompass much of human activity on planet Earth. This includes much of the land masses between temperate climates to the northern

---

[132] Lawrence Edwards, *The Vortex of Life*, 267.

[133] Lawrence Edwards, *The Vortex of Life*, 264.

and southern hemispheres, in effect the greatest areas of human occupation. Within these three systems water plays a key role which has been greatly altered in the last 100 to 200 years. Much of what we see at present as climate change is the result of this modification of the global water cycle, as we will explore at present.

The difference between full and half water cycles is crucial for an understanding of climate change. The present, disastrous management of rivers and forests, not to mention of arable land, plays a direct role in it. In what follows we will call the natural cycle of water, the 'full hydrological cycle', the one that Nature has known for millennia. More and more human activity has displaced this full cycle toward a half cycle. At the basis of this epochal contrast is once more the influence of small differences, such as moving from a positive to a negative temperature gradient. Under a purely quantitative scientific outlook the changes in the cycle of water go completely undetected. It is in fact the great contribution of Schauberger the realization that rainwater can only penetrate the ground under a positive temperature gradient, when the ground is colder than the water.

The Full Hydrological Cycle

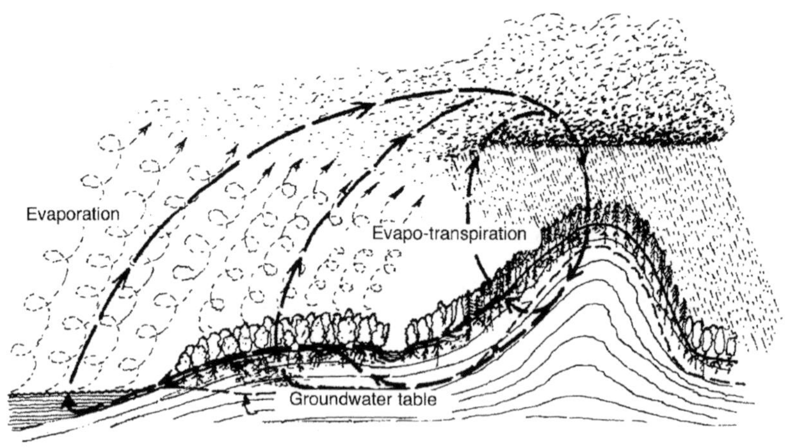

Figure 20: Full hydrological cycle

The hydrological cycle of water is the circulation that takes place between the groundwater, land (especially under the cover of forests), oceans and the atmosphere.

Water that evaporates from the great oceans' masses (3/4 of the Earth's surface) condenses and falls as rain; some sinks into the ground, the rest drains off. The ocean mainly consumes all it produces and leaves little dissolved content in the water that evaporates, producing juvenile water. Here it is that the importance of the forest cover plays a major role. The forest maintains a positive temperature gradient from the air to the soil—generally cooler than the rain and air—during much of the year, and especially in the growing season. The positive temperature gradient attracts the water towards the +4°C stratum. The absorptive capacity of the soil is thus increased. The groundwater table remains high, leading to the formation of a natural subterranean water reservoir and of springs.

Where a natural forest cover exists 85% of the rainfall is retained, of which 15% is intercepted by the vegetation and 70% feeds the groundwater aquifer. The water that sinks is enriched with minerals and trace elements. What is evaporated from the trees is of a far richer quality than what comes from the oceans under the influence of the Sun and oxygen alone. Water evaporating from the forests, carries the energetic imprint of both geospheric and atmospheric influences.

In the absence of light and heat, salts and minerals are precipitated near the boundary layer (+4°C) and present no danger for the vegetation. The water table remains high. The phenomenon of the full-cycle explains the lapse of time between rains and water discharge in the mountains. Rainfall over a few days may not show in increased river flow because the water is absorbed by the ground. The water discharges only after warmer weather arrives, causing cold groundwater to rise. Besides keeping the water table high, the restoration of the full-cycle leads to the formation of springs.

Hydrological half-cycle

The positive temperature gradients have presently been more and more transformed into negative ones. The hydrological half-cycle arises in great part where forests are felled or clearcut over large areas. Without the forest and its captured water a greater contrast arises between areas with abundant evaporation—the seas, great lakes and their proximity—and those with little or none.

If the tree cover is removed at places still surrounded by the forest, then the groundwater rises locally and problems of salinization arise. If a wider area is exposed then the groundwater table first rises and deposits salts, but then sinks because the counteracting downward water pressure has been reduced. Continuing irrigation allows a repeat of the first phase of rise and resulting salinization; it compounds the problems with repeated salt accumulation.

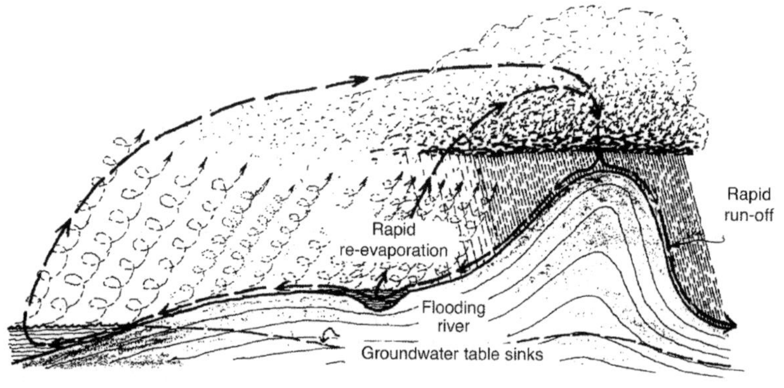

Figure 21: Hydrological half-cycle

In the absence of forest cover the temperature of the ground rises and causes a negative temperature gradient between the rain and the warmer soil. When the soil is warmer than the air, water in the ground expands at a faster rate than the falling rain, preventing it from penetrating the soil. The cool rain cannot penetrate the ground and runs off. At night the water can penetrate the soil due to the change of temperature gradient, but during the day it will bring the salts closer to the surface. A good example of this is what happens in desert areas when there is a sudden rainstorm. The riverbed fills, but instead of quenching the soil and replenishing the water table, the water runs off on the surface, possibly causing havoc.

Similar phenomena take place where the river meets the forest. When a river flows in a forested area the temperature of the water is often warmer than the surrounding soil with a resulting negative temperature gradient. The transfer of moisture, nutrients, and energy takes place from the river in the direction of the ground. The fertility of the soil is enhanced and the groundwater table recharged. When the river flows on bare ground, the negative temperature gradient prevails and the flow of moisture, nutrients and energy takes place in the reverse direction. The river extracts them from the ground, leaching minerals, trace-elements and nutrients, leading to infertility and a sinking of the groundwater. The longer a river flows through bare, irrigated farmland the more it becomes contaminated with salts, artificial fertilizers, pesticides and other pollutants.

Another immediate consequence due to the retreat of the groundwater is the lack of available nutrients coming from below to the vegetation, even the most deep-rooted trees. The trees will feed on the little that comes from impoverished, juvenile rainwater. What is more, the water that evaporates is similar to the juvenile water that fell, causing an energetic depletion or deterioration.

Floods and water run-off are coupled with the retreat of the groundwater table, which is no longer recharged. The nutrients and trace elements can no longer reach the vegetation, causing desertification. With the half-cycle there is a constant production of water vapour, and therefore rain. Already in 1930 Schauberger soberly recognized: 'One flood therefore gives rise to the next.'[134]

The hydrological half-cycle gives rise to a vicious cycle. More water evaporates and the greater vapour content in the atmosphere causes more rains. In fact the limited circulation of the half-water cycle increases the intensity of thunderstorms. A general double pattern arises: hurricanes, violent storms and flooding near the coasts, and droughts inland. Everything that the half-hydrological cycle describes is familiar at present with global climate patterns.

The reduction of forest acreage and the radical transformation of forest landscapes in artificial plantations, with the resulting hydrological half-cycle and the sinking of the water table, leads to progressive aridity in the upper reaches of the rivers—so called karstification—and the devastation of the cultivable land with the increasing development of wetlands toward the lower course of the river. On a larger scale we witness the creation of deserts in the interiors and floods and heavy downpours toward the coasts.

Modern forestry creates one vicious cycle after another. The monoculture of flat-rooted species, which cannot access nutrients other than in a very narrow band of soil is in itself very detrimental, but more challenges have been added. In traditional forestry hardwoods were harvested between 70 to 140 years, and between 15 to 40 years for softwood. At present, in the name of economic imperatives they are felled at a very immature age. Since this means a

---

[134] Viktor Schauberger, Callum Coats, editor, *The Water Wizard*, 102.

lower volume, then a larger area must be cleared, exposing the soil to a faster water cycle—a half-cycle—and the plants to unhealthy growth. Since the wood is of inferior quality it does not last long, hence the need for new wood and accelerated harvesting ... showing once more the inability of modern science to link causes to consequences.

Already in 1932 Schauberger indicated that 45 million hectares (111 million acres) of state-controlled forests in Austria had gone from natural forests to plantations in the space of 150 years.[135] Between 1928 and 1932 alone in all Central Europe the water table had sunk about 6.5 feet.[136] The sinking already showed clear consequences for the spruce, the pine and beech trees, even as far as firewood quality was concerned. Modern forestry management has led to the drying up of mountain springs—robbed of the protection of tree stands—and the desolation of high alpine pastures, which are karstifying.

The main way out of this quandary is to replant the forest, but also a real forest. Initially one may have to use salt-loving species to absorb the salts left behind. Later, after the ground is cooled off by vegetation the rainwater will take the salts to the depth, pioneer species will die off and be replaced by local species. The deforested water banks will have to be protected again. With the presence of protective species the water will recharge the water table when the soil is cooler than the river. A confirmation of this way of bringing back the original forest was found out accidentally in the community of Gaviotas in the eastern llanos—the tropical grassland plain—of Colombia, where Caribbean pines were planted over 8000 hectares (20,000 acres) which were able to survive the harsh soil conditions in a highly acidic soil. The website of Friends of Gaviotas reports: 'Over the years the pine trees have provided a shady understory for other plants and animals to thrive. Some of these species may be dormant seeds of ancient rainforests that once covered the region. The pines are slowly being crowded out by the regeneration of indigenous species.'[137]

The link between forest, water and climate was already clearly perceptible to Schauberger in the thirties. It could be summed up thus: 'The natural habitat of water is the forest. If water's habitat is

---

[135] Viktor Schauberger, Callum Coats, editor, *The Fertile Earth*, 110.

[136] Viktor Schauberger, Callum Coats, editor, *The Fertile Earth*, 96.

[137] http://friendsofgaviotas.weebly.com/forest.html

destroyed, it becomes unstable and hunts about everywhere, ceaselessly seeking out those substances needed for the growth of vegetation.' And he concludes: 'This is why the disturbance of organic processes of growth leads to the disruption of external climatic conditions. In further consequence, this inaugurates an unfavorable rearrangement in the status of the inner climate of remaining forms of vegetation, and ourselves included.'[138] In conclusion climate alteration and fate of culture are closely interrelated: 'The economic death of a people has always been preceded by the death of its forests. The latest and most terrible of misused Nature's warning signs was the drying up of springs, the increase in catastrophes and the nascent fickleness in the behaviour of water.'[139]

Modern river management, forestry and agriculture have contributed to the complete alteration of the Earth's ecosystems. What Schauberger said in relation to forestry may as well be repeated in relation to rivers and farms: 'Without exaggeration modern forestry can be described as one of the greatest threats to culture.'[140] While our present gaze is confined to the excessive use of fossil fuels in generating climatic disruption, the deeper and more pervasive drivers of the global ecological disruptions are almost completely ignored because they stem from a pervasive ignorance of the subtler forces at play in Nature, the disregard of everything of a qualitative nature.

Another hidden driver of climate change has accompanied our considerations in silent, barely hidden ways. Materialistic thinking has enshrined its tenets of complete domination of Nature and its systematic exploitation. Many of the changes that Schauberger witnessed in his time followed the economic imperatives of globalization that were ushered in by WWI. They have become the norm and accelerated after WWII and in our time. These have meant the progressive disappearance of state-owned or private forests and the accumulation of land in the hands of fewer and fewer, larger corporations. Larger and larger holdings, whether of trees or farm crops are exploited according to the factory model. Maximizing profit by standardizing operations means mechanizing and reducing manpower, hence the time for observing Nature at work and offering modulated responses to each crisis. Nothing of the sort is

---

[138] Viktor Schauberger, Callum Coats, editor, *The Fertile Earth*, 98.

[139] Viktor Schauberger, Callum Coats, editor, *The Fertile Earth*, 103.

[140] Viktor Schauberger, Callum Coats, editor, *The Fertile Earth* 56.

possible in the extensive monoculture plantations that we still call forests or farms in spite of hardly being such. Here is one of the most blaring culprits of global ecological depletion, yet one that is practically shielded from inquiry when it comes to climate change. The obvious, way too obvious, remains the elephant in the room seldom talked about.

Little work has gone in looking at local climatic alterations. The data would have to be used differently than they are at present in order to look at the phenomena to which Schauberger points. Rather than global amounts over large areas, one would have to look at regional differences in temperature and their evolutions in time, precipitation patterns such as frequencies and amounts, and distribution over the year, glacier formations, and other similar data.

Michael D'Aleo and Stephen Edelglass of SENSRI (Saratoga's Experiential Natural Science Research Institute) use the phenomenological approach to study and compare the effects on local climate of water vapour and CO_2.[141]

Comparisons between urban and rural areas lead to isolating a 'heat island effect'. An example: records of the late 1990s in Georgia show temperatures inside Atlanta up to -13°C (+8°F) higher than the countryside. And going into finer data, we see even differences between city parks and business districts, and even noticeable differences between West Point station (Harriman State Park) versus Central Park, NY.

The researchers point to the fact that when human activity burns fossil fuels it frees not only CO_2 but also water. The specific heat at around 100°C (X °F) is twice as high for water vapour versus CO_2 and four times higher for liquid water. According to what fuels we use we free in average more water than CO_2. For methane this is a ratio of 2 to 1; of 4 to 3 for propane, 9 to 8 for octane, lowering down to practically 1 to 1 for coal and diesel. In terms of energy released from gasoline combustion, this amounts 431,000 kJ for water vapour versus 40,300 kJ for CO_2, a ratio of more than 10 to 1.

In their study the authors also bring attention to the lowering of the water table and consequently water's release in the atmosphere, and to deforestation, releasing more water than CO_2, and greatly contributing to higher evapotranspiration.

---

[141] See Michael D'Aleo, and Stephen Edelglass, *Water, Energy and Global Warming* at http://www.sensri.org/global-warming.html.

While there is little net gain in precipitation amounts, the National Oceanic and Atmospheric Association's report of 1999, states that areas with longer rainy spells are balanced with those experiencing droughts. Since water does not accumulate easily in the atmosphere, it seems most of the higher atmospheric water vapour will be found around cities and deforested areas. This may explain why there is no noticeable net warming in the Midwest plains, where there has not been recent deforestation.

The above analysis is extended to the effects of human-generated energy production in big cities. Considering the per capita consumption of energy in big cities, we reach amounts along the scale of solar energy inputs. An example: energy consumption amounts to 54% of solar energy input for Queens, NY. With it goes a large amount of water released; an estimate reaches 7 million gallons of water per day in Queens. This is an amount that affects local precipitations.

The authors conclude that temperature variations worldwide are correlated with urban areas and deforestation: 'These temperature variations—including the urban-rural disparity—do correlate well with energy consumption and local water vapour production.'

We have seen in Chapter 3 how electrical energy could be generated from seawater, using the simple concept of expansion of volume, augmented by planetary motion. Before delving deeply into Schauberger's generation of etheric energy in the next chapter we can now turn to the generation of bioelectricity that requires just one further step—the use of the egg-shape—which we have seen abundantly at work in farming techniques devised by the Austrian researcher.

Another Kind of Hydro-Electric Energy

Hydro-electric power generated through hydro-turbines is utterly destructive of the structure of the water, which it renders extremely depleted and therefore aggressive on water and all forms of life with which it enters into contact. Added to this are the unintended consequences of the exposure of masses of water to the Sun and heat in artificial reservoirs, temperature pollution when the water exits the dam, dam cavitation and deterioration, etc. In the early 1920s, wanting to provide light to his forest warden's house, unconnected to the grid, Schauberger showed that energy can be produced from water, using centripetence rather than centrifugence, which would use just

10% of the water of a hydro-electric turbine for similar results, and at a fraction of the cost.[142] Needless to say, all of this without consequences for the environment.

The whole takes the form of an egg-shaped and hyperbolic cone generator. Basically it looks like an egg with its wide apex on top, narrowing toward a long vortex at the bottom. It is built out of plastic or resin and is some 3' to 5' high.

The water is fed tangentially to the surface at the top of the egg-shaped vessel through a brass or bronze nozzle, which is internally rifled in order to create a vortical, longitudinal flow—a variation that duplicates the idea of the water mains. The planetary motion reduces both pressure and friction as the water is centripetally drawn away from the sides. The water is therefore cooled, densified and energized as it goes through the walls of the egg and then through the cone at the bottom, before encountering under it a 'double-spiral, or multiple-spiral, shell-like impeller' which is connected to the shaft of a generator.[143] The entwined spirals face upwards toward the exit tip of the vortex funnel. When the water exits the cone it causes the impeller to rotate at high speed. The movement is connected to a generator that transforms it into electricity.

What has emerged from the exploration of Schauberger's discoveries so far constitutes the prerequisites upon which we can understand how energy can be generated from planetary motion and implosion, rather than technological motion and explosion. This will be the topic of the next chapter.

---

[142] More about this egg-shaped generator can be found at Callum Coats, *Living Energies*, 177–78

[143] Ibid.

Chapter 5

Ethericities and Generation of Energy

'With today's technological methods and processes the carbonic-acid sphere is ruthlessly torn from the Earth and used to generate mechanical power. Both coal and oil then contribute to the excessive build-up of decomposing energies. While the qualitative and quantitative destruction of the forest opened the cycle of self-destruction, the elimination of reserves of oil, natural gas and coal closed it.'

Viktor Schauberger

Schauberger's legacy culminated in a subject matter that projects us into the future, even the far future—the biotechnology of energy generation—and deserves attention under more than one heading. It predicates that free, non-polluting generation of energy is much more available to humanity, if we only think differently about our world and perceive forces at play which are constantly around us. It is a question that is also accompanied with the matter of moral discrimination, as the inventor discerned in his lifetime. In this chapter we will approach first the understanding of the nature of this energy, then the principles of how this energy can serve humankind. The second aspect will raise moral issues.

Our previous explorations have led us to see how certain kinds of motion, subtle differences in temperature and levitational energies, not substances or substances alone, produce life and growth. We have looked at the interplay of matter in different states, particularly the places in which the solid meets with the liquid or the gaseous state, where water meets with the gaseous state, light with warmth, etc. Such places are the roots of plants where the protoplasms mediate between liquid and solid, the sap of the trees where CO_2 in gaseous state helps the transport of water and particles in solution, the surface of the leaves and the openings of the stomata for photosynthesis and evapotranspiration, the meetings of waters of different temperatures with oxygen in rivers, log flumes or water mains, etc.

In all these places Schauberger tries to capture the miracle of life, and convey that it's always more than matter which is at play and that in fact '... the nurture of the spiritual driving force of life is always more important than the nourishment of the physical form that merely enfolds the soul. The physical form becomes a framework for the spiritual forces of regeneration once it has reached maturity.'[144] To move into a more specific direction about the 'spiritual driving force of life' the Austrian researcher coined a word that Callum Coats translates into 'ethericities'. We have met ethericities in the example of the collaboration between carbones and oxygen, or rather in the associated energies operating behind them. Now the question naturally arises: How do these ethericities translate in terms of forces and beings in spiritual-scientific terms?

Once more we find ourselves in front of a problem of language: witness the following statement among many.

> No plant is actually nourished by dissolved matter, but rather with ascended nutritive entities of geospheric provenance in a fourth-dimensional state. These diffuse ethericities can only enter the sapstream via the root protoplasms, where they are fertilized by diffuse oxygenic ethericities. The higher out-birth of this emulsion (*ur-procreation*) is an ethericity that belongs to the fifth dimension. These concentrations of dynagen emit negative hypercharged emanations in all directions and bind the positively charged ethericities entering through the skin or bark. Some of this emulsion solidifies and whatever is subsequently manifested is what we call growth.[145]

We have worked on the first part of this statement when looking at carbones and oxygen and the meeting of geospheric and atmospheric influences on water and on growth. Now we are meeting with more specific aspects of the energies, for example with the word 'dynagen'.

The latter statement reaffirms the ones that preceded it; growth is what happens at the meeting place of ethericities of different and polar nature. But what exactly are these ethericities? In this field we will do our best to find correspondences with what is known from spiritual science. Taking into account that we are entering a very fluid field of inquiry, we will hold these correspondences lightly and formulate them as hypothesis that require further deepening.

Ultimately this effort is worth pursuing because we have confirmation that all of Schauberger's ideas are more than mere speculations. Schauberger had direct perception of the realm of the elemental

---

[144] Viktor Schauberger, Callum Coats, editor, *The Fertile Earth*, 160.
[145] Viktor Schauberger, Callum Coats, editor, *The Fertile Earth*, 163.

beings, and what he saw led to immediate practical applications, not just speculations. Thus we have all of his practical work in the realm of technology—log flumes, water mains, river regulation, special dam designs, farming implements, generation of spring-quality water, etc.—and most of all the applications that generate bioelectricity or other forms of etheric energy. Even more interestingly, the boundaries between these two fields is tenuous. In generating high quality spring water in his laboratory Schauberger discovered that he had access to the generation of energy. With little change the machine designed for discontinuous use in producing water, could be used in a continuous manner to produce energy.

Two elements are here deeply interconnected in a way that flies in the face of everything we hold true in the production of energy or the working of technology. Modern generation of energy comes from a dissipative use of centrifugal motion. It generates heat, friction, noise and degrades matter, which translates into pollution. In generating energy through etheric forces Schauberger's technology offered a great output of energy, operated silently, inexpensively, and without generating pollution. On the contrary, the water that passes through these machines is of a better drinking quality at the end of the process than at the beginning. Conversely, the ennobling of air or water generates pollution-free energy. It is for this reason that the secrets Schauberger communicated and put to use assume great interest for all who want to address the ecological crisis and the matter of climate change.

Ethers and Ethericities

For all those energetic qualities detectable in the living, be it in plant growth, soil, water or air, Schauberger coined the term which Callum Coats translated as 'ethericities'. He is clearly referring to etheric energy, the energy which flows through the ethers into the physical world and back again. Schauberger fully demonstrated what Rudolf Hauschka understood from an anthroposophical view of chemistry.

By ethericities Schauberger meant catalytic, vibratory energies that work into the physical. More specifically he refers to four such forms of etheric energy and these are closely involved with water and forest, plant growth, maturation and reproduction. Among the ethericities are the fructigens (enhancing fruitfulness), oxygenes (originating in the sunlit atmosphere and promoting growth) qualigens (enhancing quality) and dynagens (enhancing immaterial energy). All ethericities are produced through planetary motion and positive temperature gradients.

Hierarchy of ethericities

The so-called fructigens can be called the 'energy aspect of carbones'; they are of a maternal, feminine nature. They unite in places with the oxygenes, energies of a masculine nature descending through the atmosphere. Oxygen is just a physical residue this energy takes. The union of carbones with oxygenes produces growth and frees the fructigens. Fructigens are the embryos of qualigens; the growth of plants can be seen as the ballast, discarded when the qualigens are radiated. The released qualigens allow the expression of life to flourish in all its multiplicity.

Finally, for Schauberger the dynagens are the most potent and refined of all ethericities. They both enhance life and set the conditions for the forming of future fructigens. We are here rising in terms of refinement of energy, while also creating a circulation loop. Let us review the ethericities one by one by gathering Schauberger's indications.

Carbones and fructigens

Carbon compounds are the building blocks of all physical forms on Earth; they combine with a great variety of elements. Decaying plant forms are either transformed in humus or in coal, oil or gas. Fructigen then is the energy potential enclosed in this store of carbon compounds. The concentrates of fatty-matter solidify after death. The fatty, emulsified substances form a contrast with the acidic quality of oxygenes.

Fructigens form the first Earth ethericity. It is energy released from the physical remains of life on Earth, as they break down in the soil. Fossil fuels are the bank reserves of future fructigens, hence their importance in the underground layers of the Earth. They are absorbed by the groundwater as it rises upward and absorbed via the root protoplasms. Fructigens are more active in cool conditions. Schauberger summarizes that fructigenic ethericities are fertilizing ethericities: 'that have entered the blood and sap vessels via the intestinal filter [skins, membranes, gills, lungs, eggshells, seed casings, protoplasms]. These originate in the Earth and are therefore predominantly negatively [geospherically] charged.'[146]

---

[146] Viktor Schauberger, Callum Coats, editor, *The Fertile Earth*, 13.

Oxygenes and Oxygen

Oxygenes are more than just O_2; they form an ethericity related to the action of the Sun. Oxygen and oxygenes become more active with light and warmth. Schauberger calls oxygen the waste product of the Sun, a solidified sunlight energy. Rainwater absorbs oxygen and becomes positively charged. The mother trees, towering over the undergrowth render this oxygen more ready to be absorbed by the underlying trees. Oxygen is also absorbed in rivers in the eddies through the action of vortexes on the water's surface.

Schauberger doesn't tire to remind us that: 'This *ur*-source of life comes into being when the ethericities of Earth bind those of Heaven. For this the maternal forces and energies must be more powerful than the incident fertilizing substances, for if the process takes place in reverse order, the fire is created.'[147] In other words the carbones need to be the most active element, binding the oxygen at cool temperatures. At higher temperatures the reverse occurs, but it promotes decay.

The oxygenes penetrate the atmosphere and descend; fructigens spiral upwards through the Earth via the groundwater. The marriage of oxygenes and fructigens produces plant growth. In most of our machinery the polar opposite reaction to the above occurs: oxygen and carbon compounds unite in a centrifugal reaction, under heat and pressure the oxygen has the upper hand, and the process is accompanied by noise and destruction of the substances. The hydrogen binds the aggressive oxygen to form water. Under technical motion, carbon and hydrogen, which form the building blocks of all life-forms, will explode and burn. With coolness and a biological vacuum their interaction leads to carbon-based vegetative growth and to water. The water resulting from this reaction is juvenile water, not yet mature because it has to descend through the Earth and rise again to acquire salts and be recharged with fructigens.

Qualigens and Quality

Schauberger points out: 'Goethe called nature's principal, formative essence the "Eternally Female" and "All-uplifting." I have named it "qualigen."'[148] And further: 'These embryos of qualigen or geospheric, negative ion concentrations are the true atoms responsible

---

[147] Viktor Schauberger, Callum Coats, editor, *Nature as Teacher*, 98.

[148] Viktor Schauberger, Callum Coats, editor, *The Energy Evolution*, 16.

for formation and levitation.'[149] Qualigens radiate outward and affect the quality of processes for all living things, from the quality of vegetative growth to the quality of human thinking, and our physical and mental health. The presence of qualigens is also what renders stimulating a simple walk in wild places. Qualigens endow all forms of growth with the natural ability to reproduce and establish themselves on a higher plane of evolution. Water deficient in qualigen becomes diseased and cancer-promoting; it gives birth to pathogenic bacteria.

If there is enough qualigen in water, gravity can be partially overcome. The elimination of the planetary motion of water—as in modern river regulation—leads to a decrease in the presence of qualigens and thus of quality. The activity of dynamization on the contrary will bring in qualigens. It is no wonder we increase and refine quality through dynamization of preparations.

Schauberger, who could perceive the ethericities and their diminishing presence, spoke at times in pessimistic terms about the future: 'Therefore, for the lack of the supply of qualigen, mental abilities will constantly deteriorate in step with the progressive rise in the quantity of living flesh. People will become imbecile, vegetative and the lower orders of life will atrophy and little by little will cease to exist, because the relatively highest energy-concentrate, the *"upwardly radiating"*, is missing, which manifests and declares itself as Life.'[150]

Schauberger indicates: 'Qualigen can only be produced with *cold systems of flow*, in which the more thoroughly atomized the substances (sediments) contained in the water become, the more powerful the life-energies.'[151] Note that it is in the increased interfaces that the elementals work. Qualigens' production can thus be increased with recourse to planetary motion, with the aid of matter in suspension or diluted in the water or air.[152] This is what we see in rhythmic shaking of preparations or homeopathic remedies, and what arises in the formation of the longitudinal vortexes in rivers and is released in the so-called 'water cannon'. Schauberger explains further:

---

[149] Viktor Schauberger, Callum Coats, editor, *The Fertile Earth*, 77.

[150] Viktor Schauberger, Callum Coats, editor, *The Energy Evolution*, 41.

[151] Viktor Schauberger, Callum Coats, editor, *Nature as Teacher*, 103.

[152] Viktor Schauberger, Callum Coats, editor, *The Fertile Earth*, 77.

If movement is triggered by traction or suction, then an increase in power and efficiency results. This is due to the cooling, reactive forces evolving during this dynamic process, which manifest themselves as levitation. Levitation is a forward and upward surge which is intensified by reactions and founded on processes of molecular revalorization. The resultant product is qualigen, which is the basis of all natural reproduction and higher evolvement, and the increase and ennoblement of all growth.[153]

Dynagens

In referring to this ethericity Schauberger claims: 'The highest forms of radiation belong to those arising from the high-grade transformation processes in the vicinity of the Sun.'[154] Dynagens can thus be understood as amplifiers of immaterial energy.

The effect of dynagens is dramatically displayed in the central updraft of a waterfall, where the vortex works in the opposite direction to the flow of water. An example of this is the energy that allows the trout to stay motionless in the centre of the river and dart upstream or uphill. Furthermore Schauberger also sees these ethericities in the Earth's axis: 'The Earth rotates in spiral space-curves about its diamagnetic axis. This axis is no axle, but a negatively potentiated concentration of dynagen, which arises through this inwinding motion. The axis itself is something akin to a hole into which seminal substances are sucked.'[155]

Finally, dynagens are the energy counterpart of metals, which we can consider the most refined physical elements naturally occurring on Earth. In Schauberger's work we find a claim that would be interesting to further explore: that through planetary motion, under the right conditions dynagen energy can be transmuted into its 'physical counterpart', metal.[156] Schauberger did in fact work at producing growth of metals in stones under negative temperature gradients, an effort which was confirmed by Dr Zuckerkandl.[157] In the example of metallic stones rubbing under water: the release of dynagen can be seen in the emission of light which is labelled triboluminescence. Finally, it is the dynagens most of all to which

---

[153] Viktor Schauberger, Callum Coats, editor, *Nature as Teacher*, 126.

[154] Viktor Schauberger, Callum Coats, editor, *The Fertile Earth*, 100.

[155] Viktor Schauberger, Callum Coats, editor, *The Energy Evolution*, 156.

[156] Quoted in Jane Cobbald, *Viktor Schauberger, a Life of Learning*, 98.

[157] Viktor Schauberger, Callum Coats, editor, *The Energy Evolution*, 23.

Schauberger referred in the powering of his biotechnology. They are like the fire element of ethericities.

From what has been said above we see the oxygenes coming from the Sun through light and air and meeting with the fructigens from the depths of the Earth in that meeting place of microorganisms in symbiosis with the roots, which Schauberger calls protoplasms. From still higher up 'in the vicinity of the Sun' but also descending to meet with air and water are the dynagens. In between, in all that signifies the transformation and maturation of the plant we find the qualigens. This is a characterization by extremes, because it is evident that we will find all of these ethericities working together as much as separately, and therefore any attempt to give a precise one-on-one correspondence for each situation would be futile.

From what has been said above, the reader may have recognized some kinship between ethericities and ethers. Therefore, to complete the exploration we will first turn to a spiritual-scientific understanding of the ethers. In essence what Schauberger calls ethericities are the working of the ethers and the activity of the elementals.

Ernst Marti has delved deeply into the matter of the ethers and helped us discern some key notions here: first of all the difference between ethers and elements. The elements permeate everything physical. They give it existence and condense the substance but do not influence the form. As far as substances, these: 'are the condensed effects of the stars, arrested in the elements'.[158] Thus there is a differentiation at the lower level between elements and substance, as there is one between ethers, form and formative forces at the upper end.

Formative forces originate from the periphery, from precise locations in the realms of the zodiac. They are higher forces which guide the ethers. And just as the elements need substance to manifest in Nature, so do the ethers clothe themselves in formative forces. 'When the astral forces stream in by way of the portal of the stars [or streamed in during the stages of planetary evolution], they stimulate the ethers and create from them the formative forces. The spiritual forces penetrate more deeply into the elements and create in them the substances.'[159]

---

[158] Ernst Marti, *The Four Ethers: Contributions to Rudolf Steiner's Science of the Ethers*, 4.

[159] Ernst Marti, *The Four Ethers*, 8.

This can be summarized as follows:

Ethers
- Etheric   formative forces
- Physical   form

Elements
- Etheric   process
- Physical   substance (process at rest)

Elements and ethers originated in pairs during the course of the cosmic development of Earth:

- At the Saturn stage, with the emergence of space, were issued warmth and fire, which are practically one, although we can differentiate them in the stream of time. 'Warmth ether is the birth of time. Fire its dying away.'[160]
- At the Sun stage the light ether was formed alongside the element of air.
- At the Moon stage were added chemical (tone) ether and the water element.
- In our present Earth stage appeared life ether and the earth element.

The later ethers are more evolved and contain within themselves the attributes of the earlier ones. Thus the life ether encompasses the other ones. The first two ethers—warmth and light—have expansive qualities and act centripetally. The other two—tone and life—have concentrating, suctional qualities; they act centrifugally. And this overall polarity appears clearly in the work of Schauberger, though it's not always completely aligned with what we know from spiritual science.

The four ethers are arranged from the Earth outward as concentric spheres of action, with the following sequence from within outwards: life ether, chemical ether—more closely associated with the geosphere—light ether and warmth ether with the atmosphere.[161] This order is never static. It is disturbed during the day by the action of the Sun and the influences of the cosmos; at night it is brought back closer to equilibrium.

With this background in mind we can hypothetically correlate what Schauberger defines as ethericities with the ethers thus: life

---

[160] Ernst Marti, *The Four Ethers*, 16.

[161] Guenther Wachsmuth, *The Etheric Formative Forces in Cosmos, Earth and Man: A Path of Investigation into the Living World*, 46.

ether and fructigens, tone ether and qualigens, light ether and oxygenes, warmth ether and dynagens. Keep in mind that what Schauberger expressed in abstractions, is what he observed and conversed with in the realms of Nature; this means the actions of spiritual beings. The ones that appear first behind physical manifestation are the elemental beings.

The Activity of Elemental Beings

The elementals were the helpers of the hierarchies in the forming of the Nature kingdoms. The elementals help the hierarchical creator powers, by becoming enchanted in various earthly forms, making their existence possible. To this purpose they sacrifice their etheric substance. The cosmic creative Word, orchestrating the work of the hierarchies, can only become manifest on Earth through the elemental beings. 'What the elemental beings have called into the world, as it were, is the last reverberation of the creative, formative, cosmic Word, which underlies all activity and all existence.'[162]

The next characterization of Rudolf Steiner addresses the work of the four classes of elementals:

> The gnomes carry the life ether into the roots; they carry the ether in which they themselves live into the roots. Undines cultivate the chemical ether in plants, sylphs the light ether, and the fire spirits the warmth ether. Then the results of the warmth ether unite with the life below. Undines carry the action of chemical ether into the plants, the sylphs the activity of the light ether, and the fire spirits the warmth ether especially at the stage of the blossom.[163]

We will find the activity of each type of elemental most strongly where a phase meets another, e.g., earth and water, water and air, and in the organs that are placed at the intersection, such as roots, leaves, blossoms. 'Certain elemental beings can most easily be seen wherever the different kingdoms of nature come in contact with one another. Such beings can take hold where moss is growing on rocks. They can become manifest where the bee contacts the flowers.'[164] Some examples of these: the gnomes are found where

---

[162] Rudolf Steiner, *Man as a Symphony of the Creative Word*, lecture of 4 November, 1923.

[163] Rudolf Steiner, *Man as a Symphony of the Creative Word*, lecture of 2 November, 1923.

[164] Rudolf Steiner, *The Festivals and Their Meaning*, lecture of 7 June, 1908.

metals are in contact with the minerals, the undines where plants touch the mineral kingdom, where plants and water and rocks touch one another. Let us look at each one of the four groups more closely.

The gnomes' body consists of life ether, which they carry into the roots of plants. They bring into movement the earth's minerals and make them accessible to the roots of plants. They work most of all in those living entities of microorganisms in symbiosis with the roots, which Schauberger calls protoplasms. The gnomes play an important part in the materialization and dematerialization of earthly substances, and in the creation and destruction of all the physical forms of the mineral, plant, animal and human kingdoms. In the step to materialization they become enchanted, and they are released with dissolution and dematerialization. A striking example of where this disenchantment occurs, and therefore energy is liberated can be experienced in the growth that results from bringing rich ground rock into the soil. In relation to this see the work of John Hamaker in increasing the depth of his topsoil in his Michigan farm from 4" to 4' over the lapse of ten years. For that purpose he used finely ground rock of igneous origin, rich in minerals, salts and trace elements. The finer the dust of crushed rock the more immediate are the results, indicating the increased activity of elementals, especially gnomes, at the interphase of solid and liquid states.[165]

The undines can be likened to world chemists. Their activity renders possible all transformation of substance. Steiner tells us that: 'they loosen and separate and bind the substances in the air, which they bring into the leaves in a mysterious way; they bring it to what the gnomes have pushed up from below'.[166] Thus they are very present where plants and minerals come together or water and air come together. We can observe their work in plants as they come out of the ground, in the leaves and before they are ready to bear fruit. Finally, undines are particularly active in springs at the confluence of water, plants and rocks.

The sylphs' consciousness lives in the airy element and their etheric body consists of spiritual light ether. They become enchanted in all forms of gas and are present in the atmospheric movements.

---

[165] Quoted in Callum Coats, *Living Energies*, 265–66.

[166] Rudolf Steiner, *Man as a Symphony of the Creative Word*, lecture of 2 November, 1923.

They carry light into the plants and permeate them with it. 'The light or the sylph-power in plants works upon the chemical forces which undines have brought into the plants, so that sylph-light and undine-chemistry interact there.... The sylphs really weave the archetypal plant in the present plant out of light and the chemical activity of the undines.'[167] Furthermore they are connected with the processes of ripening.

The salamanders' etheric body is formed of warmth ether and their consciousness lives in the fire element. They are enchanted in all forms of earthly warmth. They carry the warmth into the plant blossoms, e.g., in the interaction between stamens and ovaries. 'The ovary is the male element which is brought down from the universe with the aid of the fire spirits. ... Fructification takes place under the earth during the winter, when the seed comes into the earth and meets the forms which the gnomes have received from the work of the sylphs and undines.'[168] No generative increase would be possible without the work of the salamanders. For the seeds they maintain the proper temperature and transform lifeless warmth into living warmth. They are the preservers of the seeds and carry them from one generation of plants to another.

We have moved from ethericities to ethers and finally to the elemental beings behind which stands the activity of still higher beings, the hierarchies. The four groups of elementals together build up the etheric body of the Earth, leading Steiner to conclude: '... what the modern physicists calls laws of nature are the thoughts of these beings, who think on the physical plane and have their bodies on the astral plane. The creative nature forces are beings, and nature laws are their thoughts.'[169]

Much of what we saw from Schauberger's living perceptions in Nature is just what happens at the place of meeting between physical states—gas, liquid or solid—and the meeting places between realms—mineral with liquid in the soil, solid with sap, sap with atmosphere and light, water and air in rivers, etc. Here is where the activity of elemental beings translates into etheric energy.

---

[167] Rudolf Steiner, *Man as a Symphony of the Creative Word*, lecture of 2 November, 1923.

[168] Rudolf Steiner, *Man as a Symphony of the Creative Word*, lecture of 2 November, 1923.

[169] Rudolf Steiner, *The Foundations of Esotericism*, lecture of 30 October, 1905.

Where that activity is present most strongly is where growth or generation of etheric forces takes place. Thus, whether we look at the four ethers or at the four groups of elemental beings, we can confirm in general terms that what Schauberger perceived were the ethers and the activity of elemental beings. In general terms then the dynagens provide the etheric energy released by salamanders and in the warmth ether; oxygenes relate to sylphs and the light ether; qualigenes to undines and tone ether; fructigens to gnomes and life ether. How correct Schauberger was in each and every observation is a legitimate question which could be looked at in greater detail. Translating perception into correct interpretation and clear language is after all a great challenge, all the more so when you are limited by the strictures of a language you are creating ad hoc. An example of this is Schauberger's frequent inversion when speaking of gravitating matter from above (oxygen) and levitating matter from below (carbones).

It is the overall understanding of the etheric world and of the balance between centrifugal and centripetal forces that further allowed Schauberger to devise the machines for the production of etheric energy. Central to all of these are a number of polarities: between technological (decay processes) and planetary motion (upbuilding, levitational processes), centrifugal and centripetal forces, explosion and implosion among others. And all of this is built on the understanding of the role of temperature, particularly the use of cooling processes under negative temperature gradients.

Just as the explosion engines depend for their working on the exertion of great pressures, so do the implosion engines devised by Schauberger for energy production or matter ennoblement depend on the creation of what he called a 'biological vacuum'. We have seen in looking at the vortex and what it produces that this vacuum is not a simple rarefaction of matter.

The biological vacuum that Schauberger produced in his machines was the product of centripetal movement of air or water in which the cooling and condensing effects create pressures lower than those of a vacuum and situations in which substance is torn asunder under enormous speeds and surface tension. This is what densifies substance, transforms it and liberates the 'ethericities' that are then put to use.

Implosion and Energy Production

Schauberger reminds us: 'Implosion is no invention in the conventional sense, but rather the renaissance of ancient knowledge, lost over the course of time.'[170] Schauberger's first implosion motors came as a progressive enhancement of another devise, the 'repulsator', designed to return water to a naturally energized state and render it similar to spring water. Here we recognize some of the trademark approaches of Schauberger. The first ones are the choices of form, materials and motion. The machine used an insulated 10–litre copper egg-shaped vessel with some surfaces silver-plated. It produced alternating right and left-hand vortexes spinning the water in such a way that it turned colder and denser. During the spinning CO_2 is added, as well as various trace elements to simulate the chemical composition of spring water. The alternation of right and left-hand vortexes is reminiscent of what happens at the bends of the rivers or in the stirring of BD preparations 500 or 501. The process takes about 45 minutes and produces a water that helps restore health to the body. The added CO_2 plays an important role: through the cooling effect of planetary motion it is transformed into carbonic acid (H_2CO_3) and the change of phase contributes to the creation of a vacuum, through which the transformed physical substance releases pure, formative energy. The machine which operated in a discontinuous way, with a one-time input, could be modified and improved to operate with a continuous flow and thus generate energy.

The repulsator needs no head-water pressure, as would a water turbine. It is set in motion by a starter motor which only supplies the initial impetus for the naturally accelerating, frictionless rotation. New challenges arose when the application was used for generating energy. Rather than how to start the machine the more serious problem was how to stop the levitational forces from literally lifting it off the ground; it needed to be solidly anchored to the floor.

What we see to start with is the part played by special materials, in this instance copper and silver, and the importance of the egg-form. Copper and silver are diamagnetic materials which support levitative forces; the egg shape is a path-curve naturally occurring in Nature, a form which enhances levitation, supports

---

[170] Viktor Schauberger, *Implosion* magazine, # 36, p.3, quoted in Alick Bartholomew, *Hidden Nature*, 257.

and maintains natural movement based on the slightest positive temperature gradient. Copper and egg-form replace the more common steel and cylinder, which would have diametrically opposite results. The latter promote ferromagnetism, hinder natural flow and are detrimental to the growth of levitational forces.

The machines designed by Schauberger use the combination of both kinds of motion, technical and planetary, on a common developmental axis in rhythmical alternation, in order to produce a biological vacuum. Obviously the suctional forces must have the upper hand over the forces of pressure. The ratio of each force is important and set at around 96% suctional forces (planetary motion) to 4% pressural forces (technical motion). Only 4% of the original formative forces therefore are lost. The reverse ratio takes place in technical motion of our combustion engines. Professor Felix Ehrenhaft, who helped Schhauberger in this part of the research, figured out that the power of suction of implosion machines is some 127 times higher than that of explosion machines.[171]

In centrifugal motion levitational forces act as the resistance. In the implosion motor the reverse is true: the resistance comes from gravity. In technical motion the resistance/friction to movement increases by the square of the velocity, creating a dissipation of energy into heat. In planetary motion, using centripetence, friction is overcome and movement can sustain itself after an initial impulse of centrifugal energy. In some of Schauberger's machines this could be done mechanically with a few cranks of a manual handle.

The formation of a biological vacuum is achieved with specially designed rotating, inwinding whorl-pipes, similar to those used in the Stuttgart experiments with Professor Pöpel (see Chapter 1). The pipes have a slightly modified egg-shaped cross sections. They converge from the periphery toward a centre following a spiral movement so that the medium becomes colder and denser. To the spiralling form of the pipe is added the longitudinal vortexing motion inside the pipe, mimicking the movement occurring in the core-water of the river. By regulating the rotational speed in the specially shaped and alloyed container, like the one we saw in the case of the repulsator, a biological vacuum can be reached in proximity to the anomaly point (+4°C) and maintained almost constantly. At this temperature the oxygen becomes inactive and is bound by geospheric carbones and the fructigenic ethericities. In absence of

---

[171] Alick Bartholomew, *Hidden Nature*, 249.

friction the biocentrifugation (inwinding) accelerates the speed of rotation. In other words, the process self-accelerates. The speed of the water in absence of friction reaches up to 1,290 m/sec, or four times the speed of sound. At this speed the water comes out of the machine in a solid state according to eyewitnesses who saw it at work.[172] If, instead of water, air is used in the machine, this liquefies.

For every 1°C of cooling the volume of the gases is reduced by 1/273. In air containing water vapour the compaction of air to water amounts to 816 to 1.[173] This is the basis of implosion energy. Through the immense reduction in volume a biological vacuum is created, from which can be harnessed an energy that is environmentally pure. Through the compression of matter is freed immaterial energy. We can follow how these changes appeared to Schauberger's thinking:

> The densation is not limited to physical space alone. On the contrary, it is also raised to a spaceless condition through an increase in the frequency of the oscillations. It is this spacelessness, or vacuity, that generates the biological vacuum. The relation between the material, energetic and more subtle worlds should be perceived as a pyramid, wherein coarser, less energetic matter occupies the lower portion. As the volume reduces with height, the proportion between matter and energy gradually reverses until at the very apex all that is left is extremely fine matter or energies in a subtle or etheric state, above which the biological vacuum begins.[174]

The Austrian researcher is describing a process of potentization and dematerialization, which brings us back to the discoveries of Samuel Hahneman or Rudolf Hauschka.

Closing the circle on everything we have seen so far, it isn't surprising that we come back to the beginning. Good energy generates good water. The release of energy in Schauberger's biotechnology is accompanied with formation of juvenile water with great levitational energy, which can dissolve a great amount of salts and substances. This is no wonder, since this is what Schauberger originally produced with similar machines before he retooled them for the production of energy.

Schauberger's devises offered a great output of energy and operated silently and inexpensively; also without generating pollution.

---

[172] Alick Bartholomew, *Hidden Nature*, 254.

[173] 1 litre of water at 4°C weighs 1 kg, whereas 1 litre of air weighs 0.001226 kg.

[174] Viktor Schauberger, Callum Coats, editor, *The Fertile Earth*, 22–23.

The implosion motor imitates natural processes of growth and that is why it ennobles matter, giving rise to ethereal and energetic products. Through his implosive technology Schauberger claims: 'the entire production of power, heat, cold and light can be reduced in cost by slightly more than 90% and ... with this formative synthesizing current all growth can be increased by about 30% per annum and qualitatively improved.'[175] When this is looked at closely, what it produces is the contrary of nuclear fission (explosive); it can also be called 'cold fusion' (implosive). However, the production of such abundant and clean energy must be weighed against moral imperatives.

The Moral Question

In essence Schauberger demonstrated what Hauschka intuited in its farthest reaches. Energy is continuously transformed into matter and then again to energy. Matter derives from spirit and goes back to spirit. At the base of life, as in the case of the seed, new matter is formed from formative forces as Von Herzeele and Hauschka demonstrated. At the summit of the tree or plant, matter is transformed into fine, potentized etheric energies. When this reality is apprehended the ennoblement of matter and the release of energy will become matter-of-fact knowledge. But before this time will arise various other elements need be taken into consideration.

Because they don't have an ego the elemental beings at work in machines are easily influenced by ahrimanic powers, especially when human beings remain bound to materialistic thinking. This is the reason why the generation of etheric energy, as a matter of great responsibility, motivated Schauberger to claim: 'The implosion machines ... can only be entrusted to those who place the common good before their own well-being. They should have absolutely no interest in clinging to any kind of craving for power.'[176]

To orient ourselves in the moral question we must first differentiate between 'atomic technology' and 'resonance technology'. The first only ends with nuclear energy but otherwise includes all technologies that work over the forces of the lifeless world, in fact all technology as we know it. In this realm operate the highly intelligent but also very malevolent elemental beings of birth and death, which ardently wish to destroy humankind. 'It is they, not human

---

[175] Viktor Schauberger, Callum Coats, editor, *The Energy Evolution*, 54.

[176] Viktor Schauberger, Callum Coats, editor, *The Fertile Earth*, 34.

beings, who are behind what we are developing,' claims Paul Emberson.[177] If this were the only kind of technology at humanity's disposal in future times the ahrimanic beings would end up working against the goals of human evolution.

Rudolf Steiner coined the expression 'mechanical occultism' in relation to resonance technology. Beyond this also enters what he called 'moral technology', which will not enter our immediate considerations. An example of moral technology are the machines created by John Worrell Keely in the second half of the nineteenth century. Unbeknownst to him his machines responded to certain vibrations coming from his own etheric body. This is only possible when the operator acts selflessly. The machines could not be put to use because no other than Keely could set them into motion. The same machines would have ceased to work if egoistic people attempted to use them. Whereas atomic technology works specifically with human egoism, moral technology will be the prerogative of that part of humanity that will be able to operate selflessly in the future. Resonance technology stands a step in between atomic and moral technologies.

Before developing his moral technology Keely had devised machines which would respond to musical tones, played on an instrument, sung or hummed. His resonance technology works in two steps. The first impulse is given by the human being, which sets in motion an oscillatory system in the machine. Then follows the transformation of this initial impulse into a source of energy through sympathetic vibration.

The whole resembles in a first step the tuning of an instrument in a room where there is an already tuned instrument. When the first is tuned the second will resonate. In resonance technology the second instrument would react on the first with an amplifying effect, and the two would continue to amplify each other up to a certain limit. This causes a multiplication of force through sympathetic vibration. Obviously two violins left alone would not amplify each other. Resonant technology machines would be able to be set in motion through oscillations such as musical tones produced by a human being, which would then be amplified.

A variation of the above is contemplated by Rudolf Steiner. The premise to this is the growing ability of coming human beings at the time of Christ's reappearance in the etheric to start perceiving

---

[177] Paul Emberson, *Machines and the Human Spirit: The Golden Age of the Fifth Kingdom*, 20.

the etheric world and the life processes in human beings and in the natural world. Those who awaken the new organs of perception will know that life on Earth can only be explained not by the force of gravity, but by levity, the action of the etheric forces. 'But through this insight into nature 's rhythms one is led also to a certain practical application of rhythmic processes in technology. This is the goal of future technology; to develop a tremendous motor force by means of the resonance of oscillations—of oscillations that are small at first but would amplify each other simply by the working of resonance.'[178] We have seen that Schauberger's machines work in such a way that they self-amplify and generate perpetual movement. And from everything this book substantiated it is obvious he had strong perceptions in the realm of the etheric.

It is in the same lecture that we are led to understand the challenge to which Schauberger's life was subjected, and no doubt the reason why Steiner followed his destiny closely. In fact he tells us:

> These forms of knowledge may be bestowed upon mankind only if, simultaneously with a development towards them, there comes into being as widely as possible, in connection with our third point, an entirely selfless social order. No rhythmical techniques can be introduced without causing harm to mankind, unless at the same time a selfless social order is striven after; to an egoistic society they would bring only hurt.[179]

What Schauberger most ardently desired—communicating his knowledge about the generation of etheric energy—was also something that in effect would have worked contrary to his deeper wishes and humanity's needs. The Austrian, it seems, was protected by fate against reaching the premature spread of his technology: first in such a way that it was spared from Nazi Germany, then from possible American dangers. To some degree the technology is protected by the fact that only people like Schauberger could take this work further and come up with working models. But once that happens then the safeguards would no longer be there. In the end the inventor's discoveries are still a gift to humanity's future, though they will have to await better times and a less materialistic culture.

---

[178] Rudolf Steiner, *Three Streams in the Evolution of Mankind*, lecture 5 of 12 October, 1918.

[179] Ibid.

Chapter 6

Economies of Scarcity or of Abundance?

We have come to an end of our explorations about the nature of the Earth ecosystem in relation to climate and discovered that it is far from a closed system. If we take into account the ubiquitous effects of temperature gradients, different kinds of motion, the combined effects of form and substances, the generation of etheric energy, a whole other view of Nature emerges. This realty shows the far more prevailing effect of water over that of CO_2. It highlights the deep alteration of the water cycle that is taking place under a materialistic view of Nature and under a ruthless neo-liberal economic model. To this we have given the name of planetary depletion, claiming that the name is more accurate because it points not just to the physical, more visible, consequences of the ecological crisis, but also to the undermining of the cycle of water and consequent depletion of the Earth's etheric field. Since this moves our attention to the living dimension of Gaia—something far from accepted in scientific circles—it appears a deep cultural re-evaluation will need to precede a roadmap for change. The answer cannot come from computer models, which by definition completely exclude the qualitative element in Nature.

It is one matter to construct computer models and assume that Nature, and indeed even society and the human being, will behave accordingly. It is a whole other matter to look at Nature as a whole. This obliges us to integrate qualitative factors, as we have done so far, and penetrate with new thinking the continuous interplay of polarities of all sorts without which Nature simply isn't Nature, but a human construct, an unrealistic idea which may satisfy the computer's logic, but that alone.

With this second approach we will not be able to eliminate or master Nature's complexity and come up with models that give predictable outcomes, but we will be able to assess whether we can generate favourable outcomes that counter the seemingly irreversible changes in Nature, which have received the common overall designation of climate change. This is because we can include the qualitative element and recognize it even in quantitative measurements. Such would be the case of a holistic river regulation in which we are aiming at attaining and prolonging the length of positive temperature gradient stretches of river. With these we can predict the behaviour of water with little margin of error. We can do more than remediate, we can regenerate. Once the qualitative element has been taken into account we can turn to quantitative measurements at the commensurate scale. In the example of river regulation we will be able to measure the length of the waterway in which we can maintain a positive temperature gradient, what kinds of reservoirs are needed, at which depth to withdraw water from the reservoir in relation to the temperature downstream, etc.

If we fully penetrate the qualitative perspective of our natural world the results are twofold: on one hand we will find that the true state of affairs is even worse than what we think, and that the full truth is either ignored or carefully hidden and kept at bay, since it will hurt many established economic interests.[180] On the other hand we will find we can do much more than we are told.

Think what the impact would be of a full understanding of the effects of river regulation and modern forestry, and what enormous stakes are at play. An example would suffice. If we have understood the appalling consequences of monoculture forest plantations we can start to fathom the correlations between the decline of tree stands' quality and vitality, and overall ecosystem change, including shortened and accelerated water cycle and much lower resilience of the trees to external factors, all of which lead to local climate alterations. When news hit of devastating forest fires we will have a broader understanding of the factors at play. It is precise human choices in the realm of forestry that bring about ecological disasters. Many of the conifer single species forests are, simply put, tinderboxes exposed to and defenceless against even minimal ecological stresses. A tree that should grow protected from direct sunlight, has grown unnaturally under the conditions of clearcutting. It has protected itself with branches down to the very base. Once the new stand fills in and the trees are shaded, all the lower growth dies back and dries out, presenting an enormous mass of easily inflammable dead wood. Single species conifer plantations exploit

---

[180] A good way to put in words the initial experience of reading Schauberger is that of Wilhelm Balters: 'You may have lived a calm and contented life—but from the moment you come face to face with the ideas of Viktor Schauberger, you will never again have peace in your soul.' (Quoted in Olof Alexandersson, *Living Water*, 15.)

only a very narrow superficial layer of available water. Under the conditions of new saplings exposed to the Sun, the groundwater will inevitably recede, and the tree will only grow on rainwater and the scant nutrients it contains. A truly diverse natural forest would be otherwise resilient. When we hear of forest fires and know about the trees' natural behaviour we will not be satisfied with an abstract and vague claim of climate change that can justify anything and everything without showing clear correlated dynamics.

At present climate change has gone from being a consequence of human choices to being used in our reasoning as a driving cause. Our actions, we are told, primarily at the energy level, have caused climate change. Now, every time we see an ecological problem, climate change is invoked as the cause in lieu of the human mis-management that could offer a much more immediate, convincing and adequate explanation on the ground. At times climate change becomes a blanket statement and easy certainty in the news. More rains, floods, fires, draughts, famine, human disasters? Don't think too hard; it's because of climate change!

Though this work advocates primarily changes in our scientific and cultural outlooks, it is encouraging to notice that existing and extensive initiatives within the technological and economic contexts, exhaustively outline that it is possible and necessary to question the impact of our present global economy on climate. Not only that; it is also possible to show how to reverse the course, as it has been rendered explicit by the proposition of the so-called 'Blue Economy'. Not surprisingly this new paradigm rests on something very similar to Schauberger's '*Kapieren* and *Copieren*' through industrial processes that mimic natural processes using low temperature, low pressure and low energy inputs.

We will now turn to two radically different sets of solutions to the climate crisis: one within the prevalent view generated by climate models and their logic, the other turning this logic radically upside down.

Red and Blue Economies

Converging toward a more organic thinking new developments in combatting climate change highlight how much humanity finds itself at an epochal divide in its understanding of Nature. On one hand economics that honours Nature's organic complexity, on the other one which reduces it to a mechanism. On one hand, Nature can only be understood if we stretch and transform our thinking in order to grasp the workings of polarities in holistic units of greater and greater complexity. At the other end, the issue can be simplified to the extreme and become merely technical, as we have seen in computer models and as we will see once more through an example.

In Steiner's words in the last of his letters to members: 'By far the greater part of that which works in modern civilization through technical Science and Industry—wherein the life of man is so intensively interwoven—is *not Nature at all, but Sub-Nature*. It is a world which emancipates itself from Nature—emancipates itself in a downward direction.'[181] It is through the lens of sub-Nature that our planet Earth can be seen as a purely physical mechanism and a closed system.

What is true in the scientific realm is also visible in economic models. The dichotomy Nature / sub-Nature can be discerned clearly if we penetrate two distinct views of the economic response to climate change. They are clearly differentiated by their respective technologies.

Sub-Nature technological solutions go hand in hand with a sub-Nature economic system, which emphasizes large scale and centralization, bringing about systemic waste of resources and energy, environmental degradation, economic disparities and marginalization of whole sectors of the global population. A more natural economic approach goes hand in hand with decentralization and the creation of efficient economic ecosystems that save energy and resources, and meet the needs of people and ecosystems.

We touch on this matter here because, while this is not the object of this book, leaving it completely aside would be like ignoring another big elephant in the room. Our economic system is the root engine of what we see as climate change, or planetary depletion to use the language of this work. There is yet another reason for adding this theme here. Schauberger has helped us to see countless technological innovations that can be used at a small scale, that ennoble matter and energy in the process. They are eminently fitted for a decentralized economy.

The question of how more naturalistic innovations would integrate a holistic economic system would have been a purely

---

[181] Rudolf Steiner, letter to the members *From Nature to Sub-Nature* of 25 March, 1925 in *Anthroposophical Leading Thoughts*.

academic question until I approached the work of Gunter Pauli and ZERI (Zero Emissions Research and Initiatives) and the economic model they propose. We will outline here some of its hallmarks and send the interested reader to the Appendix for the larger picture that is possible from its premises. First we turn to new developments in the fight against climate change.

Mega-Facilities for Trapping CO_2

The world's largest carbon removal project, situated in Wyoming, speaks volumes for many similar attempts to purely technological solutions to the climate crisis. It is the so-called 'Project Bison', the brainchild of the Los Angeles-based Carbon Capture Inc. and Texas-based Frontier Carbon Solutions.[182] This is not the first such operation, only one on a much larger scale, than the largest such plant already operating, the Icelander Orca, which captures some 4,000 tons of CO_2 per year.

Climeworks, the Swiss company, responsible for the Orca project announced a 10–year deal with Microsoft, which will allow the tech giant to offset 10,000 tons of its CO_2 emissions. For this purpose Microsoft tapped into its $ 1 billion Climate Innovation Fund.[183]

The Orca plant pales in comparison to the ambitions of Bison. The initial 12,000 tons of CO_2/year will accrue to an estimated 200,000 tons by 2026, and 5 million tons by 2030. The facility will be comprised of large arrays of modules, the equivalent of 40–foot shipping containers. The technology makes recourse to 'reactors' acting as filters that extract CO_2 from the air. The gas is then pumped deep underground to be permanently stored in saline aquifers. Notice in passing that CO_2 will be needed in order to dispose of … CO_2. Notwithstanding the lack of studies for the unintended consequences of such an operation, the sheer ambition of the undertaking should offer pause to think.

Adrian Corless, CEO of Carbon Capture, waxes ecstatic about the business possibilities: 'So even with this project, we're still in the early days of what needs to become an industry about as big

---

[182] Joseph Guzman. The world's largest carbon removal project will break ground in Wyoming, 1 Oct. 2022, available at https://thehill.com/changing-america/sustainability/climate-change/3669378–the-worlds-largest-carbon-removal-project-will-break-ground-in-wyoming/.

[183] See https://www.geekwire.com/2022/heres-a-look-at-the-worlds-biggest-carbon-capture-site-which-just-signed-a-10–year-deal-with-microsoft/

as the oil and gas industry is today.' His are not empty words since in fact the support is there at the political and legislative levels. The Biden Administration's Inflation Reduction Act significantly increased the tax credits for carbon removal from \$50/ton to \$180/ton. And the even more recent Bipartisan Infrastructure Law allocated \$3.5 billion to establish four similar projects for large-scale CO_2 removal.

Project Bison will be soon followed by larger endeavours, such as Stratos, the world's largest Direct Air Capture (DAC) facility, planned in Ector County, Texas. Stratos will be the joint venture of Occidental (short for Occidental Petroleum Incorporated) and BlackRock, one of the largest multinational asset managers. BlackRock will invest \$550 million in the initiative. Occidental is a multinational energy corporation based primarily in the United States, the Middle East and North Africa, and one leading producer in offshore Gulf of Mexico oil.[184]

Stratos is designed to capture 500,000 tonnes of CO_2 per year. Vicki Hollub, president and CEO of Occidental declared: 'We are excited to partner with BlackRock on this transformative facility that will provide a solution to help the world reach net zero.' And she has reasons to be excited since she intends to direct Occidental to support over one thousand similar projects by selling the very lucrative direct air capture (DAC) technology and design expertise.[185]

That this will become the landscape of our future is very clear from the support of the autoritative World Economic Forum (WEF) and International Energy Agency. The WEF indicates: 'We cannot reach net zero without carbon dioxide removal (CDR) technologies—they are essential to delivering the "net" in net zero. Of these technologies, direct air capture (DAC) has significant advantages over other CDR approaches.'[186] The IEA's 2022 Direct Air Capture Report adds an idea of the dimension of such projects: 'In the IEA Net Zero Emissions by 2050 Scenario, direct air capture technologies capture more than 85 Mt of CO_2 in 2030 and around 980 $MtCO_2$

---

[184] See https://www.oxy.com/news/news-releases/occidental-and-black-rock-form-joint-venture-to-develop-stratos-the-worlds-largest-direct-air-capture-plant/

[185] See https://www.reuters.com/business/energy/occidental-ceo-sees-potential-license-1000–carbon-capture-plants-2023–11–08/

[186] See https://www.weforum.org/agenda/2023/08/how-to-get-direct-air-capture-under-150–per-ton-to-meet-net-zero-goals/

in 2050, requiring a large and accelerated scale-up from almost 0.01 MtCO2 today.'[187]

Carbon credits are sold in so-called 'carbon markets', self-regulated by non-governmental entities. Among the largest players in this market are megacorporations such as Microsoft, UBS, Airbus, Shopify, Swiss Re, and newly created 'demand aggregators' like Frontier—a joint initiative of Stripe, Alphabet, Meta and McKinsey, among others. Altogether they committed to buy $ 1 billion worth of carbon credits.

The view from sub-Nature illustrated above exalts the perspective of the Earth as a closed system, the idea of an input-output model, that corresponds to certain physical and mathematical constraints, which we can direct and operate within a technologically monitored system. The above clearly illustrates how complex problems are being tackled with deridingly simple solutions that do not come from a real understanding of Nature. We can soothe our anxiety and consciences with the illusion of quick fixes, or come to accept that real change can only come from a restoration of Nature's fallen state.

A Blue Economy in Defence of the Planet

The examples given above form part of the view that Gunter Pauli calls the 'Red Economy', in contrast to the Green Economy and Blue Economy. The Red Economy: 'borrows—from nature, from humanity, from the Commons of all—with no thought of repayment beyond postponement to the future. Insatiable economies of scale callously search for ever lower marginal costs for additional unit manufactured, making dismissive abstraction of all unintended consequences.'[188] The prevailing global economy delivers a growing host of products aiming at satisfying a variety of artificially stimulated desires—through massive advertising—instead of solely addressing real needs. It then favours the growth of monopolistic corporations geared at creating economies of scale. This is the path that inexorably leads the global ecosystem toward growing pollution, environmental degradation, depletion of non-renewable resources, impressive generation of waste, loss of biodiversity and runaway carbon emissions.

---

[187] See https://www.iea.org/reports/direct-air-capture-2022

[188] Gunter Pauli, *The Blue Economy: 10 Years, 100 Innovations, 100 Million Jobs.* Report to the Club of Rome, (xxix)

A big challenge of the present global paradigm is waste, a really large proportion of everything that is produced in the present economic system. A few examples will suffice given their magnitude. Corn is only used for the seed; the process of brewing uses only the starch from barley and the rest is waste; coffee is only exploited for the berry. The production of sugar from cane uses only 17% of the plant, while the rest is burned—a systemic inbuilt and persistent contribution to the growth of CO_2.

Coffee is the second most widely traded global commodity and one of the most wasteful farming practices. It generates waste at the source, the so-called 'pulp' and a second post-consumer waste of grounds. This means that only 0.2% is enjoyed by the consumer, the rest is waste. The 2008 coffee consumption equalled 134 million bags (60 kgs/bag), which means 23.5 million tons go unused, just wasting.[189] After years of study Nestlé chose to burn coffee waste, which is 80% water, as its best environmental option, another energy-inefficient proposition and another contribution to global CO_2 levels. As if the above did not speak volumes, just keep in mind that the farmer hardly makes 0.1 cent on average out of $ 3.00 of what is charged to the customer for an espresso.[190]

The litany of waste in our economic system is endless. Paper manufacture gets rid of everything but the small amount of cellulose. The rest, the 'black liquor', goes once more out in smoke and CO_2, since it is incinerated. Looking at the global system, just the waste generated by mining dwarfs municipal solid waste by a factor of 71. Consider that in the US alone the cost of transporting waste to the landfill adds up to $ 50 billion. And this is nothing in comparison to the costs for collecting, hauling and disposing of the wastes from construction, agriculture, mining and industry, which reach a stunning $ 1 trillion.[191]

All in all we have an economic model built upon the premises of scarcity and leading to scarcity as its conclusion. It seems that in the present corrective measures for climate change we must control access to our resources and manage in the best possible way … the little that survives colossal waste. We must control the emissions from inefficient processes without questioning the very source of the problem.

---

[189] Gunter Pauli, *The Blue Economy*, 86.

[190] Gunter Pauli, *The Blue Economy*, 88.

[191] Gunter Pauli, *The Blue Economy*, 6.

Why not change the premise of the inevitability of waste of our present economic model? This is exactly what Gunter Pauli and ZERI, plus a multitude of economic agents, are attempting worldwide. Once more, brilliant solutions to the climate crisis are hidden in plain sight, but they are not seen and acknowledged because they correspond to a new paradigm, that of the 'Blue Economy'. Blue stands for the colour of the planet seen from space, and as a further differentiation from so-called 'Green Economy'. We will see shortly that Blue Economy approaches are not only good for human well-being; they address climate change upstream, rather than remediating downstream.

The short-sighted thinking of the 'Green Economy'—laudable as the intentions have been—has led to tinkering within the predominant economic model. Gunter Pauli was the founder of the well-known Ecover company, which produced palm oil-based biodegradable cleaning products. He was dismayed to realize over the years that the incentive to produce palm oil caused the destruction of vast swathes of rainforest. He painfully awoke to the drawbacks of the Green Economy and came to realize that biodegradability and renewability were pitted against sustainability in the market dynamics of the global economy. Add to this that the Green Economy has generally meant higher investment rates and higher costs to consumers. It is well-known that biofuels, besides using corn inefficiently, are driving up the cost of the staple upon which many depend.

Gunter Pauli has been a member of the Club of Rome for three decades. After creating half a dozen companies by the 1980s, he started a publishing company to render available the Worldwatch *State of the World* and *Vital Signs* to the European business community. Pauli's earlier disappointments gradually brought him to approaching an economy that eliminates the concept of waste and acts like a natural ecosystem.

In Nature nothing is waste; everything left behind by one kind of organism becomes the substratum for another organism. The realization spurred Pauli to undertake a three-year research project in cooperation with the United Nations Development Program (UNEP). It led to the founding of the ZERI foundation in Switzerland to spearhead the study of economic systems without waste and emissions, which generate jobs and social capital while being affordable. As part of the project he created an inventory of cutting-edge innovations inspired by natural organisms and systems. Out of a list of 340 technological innovations that met sustainability criteria he finally narrowed down to the 100 already enacted and feasible, or those that held the most promise and likelihood of success. The guiding criteria for selection were the ability to use efficiently nutrients and energy, moving them from one level of the economic system to another, leaving little to no waste. The viability of these new technologies has been recognized by the executive director of UNEP, Achim Steiner thus: 'Many technologies are in commercial use. We are not talking about theory anymore; these are real results occurring in the real world and in the real market.'[192]

Let us get closer to Pauli's vision. The Blue Economy purports to apply the achievement of ecosystems to economic systems. Moreover it is a 'new economic model that is not only capable of responding to the needs of all but converts the artificial construct called "scarcity" into a sense of sufficiency and even of abundance'.[193] The focus lies not just on the finished product. The processes themselves must be energy-efficient and free of waste. Herein lies the major difference with the Green Economy. Notice in passing that the above is completely consonant with Schauberger's technological applications and a phenomenological understanding of biological systems.

The Blue Economy moves past the need to replace something toxic or energy consuming with something less toxic or less energy consuming. It moves toward altogether more efficient and affordable processes. Many of the technologies used to strengthen the Blue Economy take their departure from imitating unusual processes in the natural world. Vortex technology is a very general one. Others take their departure from how animals achieve adaptation to extreme conditions and overcome natural obstacles with simplicity and little expense of energy, whereas we humans, confronting the same situation, have thus far expended great amounts of resources and energy.

Much of the Blue Economy rests on observations through which we can realize how minute shifts in pressure, temperature, and moisture content create outstanding products and processes. Instead of having recourse to the sub-Nature of oil and coal biochemistry or manipulating the biology of life, we can imitate the ways in which

---

[192] Gunter Pauli, *The Blue Economy*, 43.

[193] Gunter Pauli, *The Blue Economy*, xxx.

Nature uses physics. Nature itself, when we mimic it, is showing us ways to eliminate energy-inefficient synthetic chemistry. A great number of new technologies, in addition to using minute shifts in pressure, temperature, most often just take advantage of gravity. Energy savings as well as reduced emissions are obvious if we can shift from high temperatures and high pressures of technological motion to processes using low temperatures and low pressures, whether this be with technological or planetary motion. In fact the Blue Economy is a great intermediary step in the direction of Viktor Schauberger's technical applications resting on planetary motion. Both are based on the tenets of '*Kapieren* and *Copieren*.'

What renders the Blue Economy an ecological economy is the logic of 'cascading nutrients and energy'. The term cascading comes from the analogy with a waterfall, indicating an effortless movement from one level to the next, akin to gravity. An example is what happens in Nature where each realm feeds the next one. The minerals feed the plants; the plants the animals, and the waste of one is the nourishment of the other, closing the loop when dead plants and animals return to the soil. Industrial processes can now be designed to emulate this process of cascading. The Blue Economy can boldly claim that it 'initiates a generative and regenerative cascade of implementable innovations leading to a sustainable product, sustainable manufacture and sustainable whole systems in order to create competitive products, competitive processes and competitive business models that go far beyond core business practice'.[194]

In this alternative view of economics the ability to offer better and more diversified returns reduces the element of risk. An example: there are thirty-seven known commercial applications for the vortex. The company Watreco can help save energy in ice-making, speed percolation in golf greens, descale pipes, pump air into fish tanks among other things. Since it can produce multiple cash flows through multiple applications it helps reduce risk and increase the value of the intellectual property, raising interest in a variety of investors seeking differentiated outcomes. At times the Blue Economy offers the possibility of replacing high expenses of materials and energy purely and simply with nothing or with something infinitesimally smaller and completely different from the system it replaces, as we will see below.

---

[194] Gunter Pauli, *The Blue Economy*, 74.

Looking at solutions to the energy and climate crises we can be truly inspired by how much the Blue Economy has to offer, both by reducing energy consumption and by taking advantage of CO_2 itself.

Saving Energy and Turning CO_2 into a Resource

The new biotechnology of the Blue Economy envisions opportunities where traditional economy only sees limitations. We'll start with an example from a lowly insect, which most often society wants to exterminate. Who could fathom how much we have to learn from it? The termite is a highly skilled engineer. Its sophisticated mound architecture is such that it keeps a constant temperature of 86°F and a constant humidity of 61%.

Another source of inspiration in temperature control and ventilation is the deceptively simple zebra mechanism. The savannah animal takes care of her body temperature by creating wind micro-currents derived from the temperature differential between black (absorbing) and white (reflecting) skin surfaces. Higher air pressure moves from the white to the black areas allowing the zebra to reduce surface temperature by 17.5°F.

Using the termite and zebra's wisdom Anders Nyquist developed mathematical formulas to minimize or eliminate the need for heating or cooling systems through managing the right airflows. Examples of buildings using this technology are Daiwa House in Sendai, Japan, Las Gaviotas' hospital in Colombia, the Eastgate Centre shopping complex in Harare, Zimbabwe, the Laggarberg School in Timrå, Sweden. Several of these buildings completely eliminated the need for heating and cooling systems, others complemented them with the use of heat pumps.[195]

Elsewhere Young-Suk Shu and Tae-Sung Oh from Korea put to good use the wisdom of ants and termites, whose underground storage of plant debris, not only creates a natural compost for plants but warms the soil, thus protecting plants against freeze.[196] The two researchers transposed this idea to a greenhouse whose heating principle is to heat the roots, reducing or eliminating altogether the need to heat the air, and saving energy consumption by two thirds or more. In Japan the technology has been adopted by tomato and strawberry growers. It can now further replace in unexpensive fashion the need for radiant floor heating.

---

[195] Gunter Pauli, *The Blue Economy*, 279.
[196] Gunter Pauli, *The Blue Economy*, 280.

With new kinds of thinking the energy question can be placed in another context: it can be viewed as an opportunity. Can we be audacious enough to apply this new thinking to CO_2 itself? Could we possibly see it as a resource rather than a waste? The Swedish MRD Construction Company receives unusable recycled glass and processes it turning it into glass foam blocks used as structural building materials—at present in Belgium and Czech Republic. All that is added to the glass is CO_2.[197] The resulting product is lightweight, resistant to acids and mould and a good insulating material. It can be used instead of another four products in the market.

Canadians Normand Voyer and Sylvie Gauthier's innovation is a process that uses enzymes to capture carbonic gases and deliver CO_2 to produce calcium carbonate. Their venture, aptly named CO_2 Solutions, trades in the Toronto Stock Exchange. It is very adaptable since the raw materials can even be tapped from coal-fired plants' smokestacks. Not only does it tackle the CO_2 problem at the source, it also reduces the need to mine calcium carbonate to manufacture cement.[198] Think of the further potential when you consider that the largest CO_2 emitters are cement factories themselves, upward of 10,000 of them globally. A variation of the above is the idea developed by Geoffrey Coates of Cornell University to convert CO_2 and CO to plastics and chemicals with the help of an enzyme that transforms CO_2 into polymers.[199]

Carbon dioxide could even become a source of energy with the help of algae due to their fast growth rate. The Minnesota Center for Biorefining estimated that algae produce up to 5,000 gallons of biofuel/acre/year, which compares favourably with 18 gallons for corn, 48 gallons for soybeans and even with the 635 gallons from palm trees.[200] Algae do not detract from an existing food source by driving its price up; on the contrary they generate multiple lines of value.

In Brazil this knowledge has been put to use in the cultivation of spirulina, yielding 2000 gallons/acre/year. Easily satisfied algae can be grown on marginal land in salt water and can even take advantage of the CO_2 from the retention basins of coal-fired power stations.[201] Sequestering CO_2 while producing energy is an amazing trade off.

---

197 Gunter Pauli, *The Blue Economy*, 210.

198 Gunter Pauli, *The Blue Economy*, 280.

199 Gunter Pauli, *The Blue Economy*, 175.

200 Gunter Pauli, *The Blue Economy*, 172.

201 Gunter Pauli, *The Blue Economy*, 172.

The idea has now spread to different parts of the world. In Colorado the National Renewal Energy Laboratory has tested three hundred algae that could produce biodiesel in the New Mexico deserts. On its side the Federal University of Rio Grande in southern Brazil set up a project geared toward food production at Mangueira Lake, one of the most alkaline lakes in the world, at Laguna Morin and in rice paddies in the area bordering with Uruguay. The harvested super blue-green algae serve as nutritional supplemental food for the local population at risk of malnutrition. Part of the yield will go to produce biodiesel. The algal membranes, if isolated, are formed of esters which can be turned into polyesters without need of sulfuric acid. As an ultimate example of a cascading process, after extracting food, biodiesel and esters, what is left can be converted into ethanol.[202] The whole cascading process amounts to what the Brazilian team likes to call a 'whole photo-biorefinery'.

Within an economy that uses energy efficiently, reduces the need for entropy-dissipating processes, gets rid of waste, the perspective of combatting climate change effectively can be viewed with optimism. It is not far-fetched to claim that the Blue Economy: 'can provide a solid rationale for implementing the agenda of the Convention on Biological Diversity and the missions of organizations like UNEP and IUCN' as claims Achim Steiner, executive director of UNEP.[203] It is truly a wonder to see what the Blue Economy could achieve in addressing official climate change goals, since it addresses problems efficiently at the source, rather than downstream, as we see practically everywhere. It is all the more astounding to realize how little it is considered, publicized and advocated by global agencies or those whose voices speak loudest about climate change.

Because there is so much more that the Blue Economy can offer to addressing the issue of climate change we have added extra information in the Appendix. You will see not only a host of processes to regenerate the Earth, save energy, reduce and reuse waste. You will also discover that a whole island has been undergoing a radical process of economic change over more than twenty years, which can give us a taste of how the crisis of climate change can be turned around within a highly decentralized and participatory economy.

---

[202] Gunter Pauli, *The Blue Economy*, 174.

[203] Gunter Pauli, *The Blue Economy*, xviii.

In Need of a Cultural Revolution

The view from sub-Nature that we saw earlier on exalts the perspective of the Earth as a closed system, the idea of an input-output model that corresponds to certain physical and mathematical constraints, which we can direct and operate within a technologically-monitored system. Complex problems are tackled with deridingly simple solutions that do not come from a real understanding of Nature. We can soothe our anxiety and consciences with the illusion of quick fixes, or come to accept that real change can only come from a restoration of Nature's fallen state.

The dominating economic model is based on a perpetual flight forward. While we are generating worse and worse ecological, economic and financial crises we seem to wager that purely technical alternatives will appear on the horizon and offer stopgap solutions. While we are squandering resources, energy and whole ecosystems we have to predicate restraint on the global citizenry. Waste on one hand leads to scarcity on the other, and with it comes growing control mechanisms.

On the other hand Gunter Pauli has showed us that even in the economy we can think holistically and learn from Nature what is needed in order to return the Earth to its real state of equilibrium, one in which technology and economic activity insert themselves in a whole new human-made ecosystem that mimics natural ecosystems. Eliminating our colossal output of waste and using moderate energy inputs for processes at near ambient temperatures using low pressure will certainly address the problem of climate change at its source. So why do we keep this healthy thinking at arm's length?

Under the light of what we have discovered, coming to know the work of Schauberger has been sobering to say the least. On the other hand, we can understand from the Austrian genius that much is possible if we only think an octave higher. And we have seen what kinds of solutions are possible through a new economy, and how vastly they differ from any possible Bison or Stratos projects. Much of the path humanity is engaged upon is actually reversible. However, this requires a change of thinking and, I would argue, a tidal wave of cultural change.

We have come to the end of the exploration from the ground up, as it were, from the perspective of Gaia as a living being. The work of Schauberger and a more Goethean scientific understanding of

Nature has allowed us to ascertain what is the essence of today's massive ecological crisis. It has placed water, not CO_2, at the centre of the issue. But the matter of climate hinges on much more than life on Earth. As we will see shortly planet Earth is not a closed system in relation to the Sun and solar system either. We must, as it were, consider a much larger ecology, a planetary/cosmic ecology, which has a paramount influence on climate.

PART II

COMPUTER MODELS OR PLANETARY HARMONICS?

THE VIEW FROM THE SOLAR SYSTEM

Chapter 7

From the Oceans to the Cosmos:
The Modern Ecology Perspective

'My own personal view is that the sun's variability on the very long times-cales of millennia and over ice ages will one day prove to have a galactic component.'

Peter Taylor

Much of what will be brought forth in this chapter comes from the extensive gathering of scientific information achieved by Peter Taylor in relation to the larger question of climate and the influences that underlie it. The value of his compendium research is to give voice to a coherent body of work which has not been given its due, though it has accumulated ample evidence, which now lies well beyond the purview of hypotheses and sheer pioneer work. It has legs.

Peter Taylor has worked for over thirty years as a professional ecologist involved in major policy issues: offering critical review, promoting advocacy, advising strategies that minimize impacts, offering advice on policy or legal reform. From 1976 to 1996 he was involved in the UK's 'Alternative Energy Strategies'. Part of that job involved working with global circulation models of atmosphere and oceans—the forerunners of climate change models—in order to understand the dispersal of radioactive pollutants. Over his long career he worked both for governmental and non-governmental organizations as a legal advocate and intervener on behalf of commissions, parliaments as well as United Nations' international conventions. For twelve years he was a chief advocate and science advisor to Greenpeace International, which led him in the late 80s, early 90s to offer consulting services to the UN's Maritime Organization in matters concerning legislation to better protect the oceans.

This chapter forms a counterpart to Chapter 1. There we looked at climate models. Part of their limitation lies in historical circumstances. The period 1980 to 2000 was a period of increased warming upon which many of the premises and assumptions of models have been built. In parallel to these more and more work arose at the periphery of the scientific world—the periphery of the greatest investments of resources—which explored new phenomena that completely enlarge the temporal perspective on climate.

Expanding the time frame means moving from solely using instrumental data—the ones that models mostly use—to adding non-instrumental or 'proxy' data. Though instrumental data are more exact and can be more easily treated statistically, they only cover up to 150 years. With this record alone we miss knowledge of the larger cycles and how the last two centuries play out in relation to the past. When you refer to instrumental data alone you can ignore the knowledge of larger cycles, and attribute even large variances to a hypothetical 'natural variability'.

Non-instrumental data, or proxies, cannot calculate exact temperatures but through them patterns become more apparent. Among these proxies we count: tree-rings, deep boreholes measuring sediments' thickness, ratio of oxygen isotopes in ice crystals at the North and South Poles, or in cave stalagmites. The work of gathering these data is complex and tedious and needs different forms of statistical treatment than the instrumental data. However, the research has been extensive and the results conclusive.

To peek forward in our exploration, if we look only at the last 150 years of the instrumental data the recent warming period appears highly unusual. This is what was presented by the IPCC's Third Assessment Report of 2001. In the years that followed (2001 to 2007) this approach was heavily criticized. The IPCC itself now admits the evidence of greater variability in the past than in the last 150 years.

The View from the Oceans

Even from instrumental data alone ocean cycles and oscillations can be detected. Much of the data of the instrumental era comes from the Satellite Cloud Climatology Project, which on its own already provides enough evidence for cycles accounting for global warming. The interplay of cycles creates time lags, destructive or constructive interference according to how they interweave with each other. Let us look at these dynamics more closely.

The atmosphere stores very little of the extra heat of incoming short-wave radiation (visible light). Most of this storage takes place in the oceans and is modified by ocean currents, plus evaporation, condensation into clouds and precipitation as rain or snow.

Even within the record of a century the influence of cycles can be discerned. Among these:

- ENSO (El Niño Southern Oscillation) an irregular cycle of 4–5 years average, which had an influence in bringing temperatures to a record high in 1998. The Southern Oscillation is the atmospheric component of the El Niño current. El Niño is followed by La Niña years, often accompanied by torrential rains and floods. And La Niña has proved elusive to predict. Low ENSO activity is associated with falls in solar magnetic activity or peaks in solar output. Due to the ENSO alone the Andes have an impressive record of climate change over the civilizations of millennia, much more impressive than what we see at present, of which more below.
- PDO (Pacific Decadal Oscillation) on an average 30–year cycle. It brought temperatures down in 2007 as it entered a 'negative' phase. Since around 1700 ENSO was enhanced when in phase with the PDO. This led to a cooling in 1947, with a shift back to warming in 1977. The connection between PDO and global temperature patterns is now documented: such is a direct correlation with two warm phases, those from 1925 to 1945 and of 1977 to 2006.[204]
- NAO (North Atlantic Oscillation) which shifts from low pressure over Iceland and low to high pressure over the Azores on a rhythm of 10–20 years. During the cycle surface waters warm up or cool off. There are also low frequency oscillations of about 60 years leading to peak temperatures.[205]
- AO (Arctic Oscillation) on a 60–70–year cycle. The AO is linked to increased temperatures during its 1940s peak.[206]

The above interlinked cycles peaked in the late twentieth and early twenty-first century.

To apprehend the magnitude of the influence of El Niño events on climate, suffice here to consider Peruvian history. We could assert that the Andes, especially Peru, have known drastic climate change in the last millennia, with catastrophes of greater

---

[204] Peter Taylor, *Chill: A Reassessment of Global Warming Theory*, 112, 114.

[205] Peter Taylor, *Chill*, 102–03.

[206] Peter Taylor, *Chill*, 50, 53.

magnitude than in any other part of the world. The following are some examples of natural catastrophes and their frequency.

- 'El Niño' major crises occurred in the years 511–12, 546, 576, 600, 610, 612, 650, and 681 AD.[207]
- W. Quinn reports 122 'El Niño' events in the last five hundred years, subdivided as follows: 67 moderate (greater than 2°C variation); 45 strong (greater than 3.1°C); 10 exceptional (4°C or more) according to paleontological, archeological and written sources.[208] Ten exceptional events in Peru were recorded in the years 1578–79, 1720, 1728, 1791, 1828, 1877–78, 1891, 1925–26, 1982–83 and 1997–98. In 1891, along the northern Peruvian coast, the rains in Piura lasted more than sixty days; the Piura River reached a width of 150 m and a depth of up to 7 m deep, compared to 30 m and 1 m respectively in normal times.[209] In addition to El Niño, when La Niña hits, the ocean surface temperature decreases. This is accompanied by rising waters in the Western Pacific. From 1958 to the present there have been twelve such episodes, accompanied by abnormally low precipitation in Piura, northern Peru.[210]
- Climate fluctuations over the centuries have been such that they have affected Lake Titicaca in a dramatic way. Around 8000 BC., it is estimated that the water level dropped by 50–60 metres. In the years 5000–2500 BC., the waters rose again. In 1000 AD they were still five metres higher than today.[211]
- In the glaciers of the mountains surrounding the high plateau there is evidence of periods of drought and large periods of rainfall between 650 to 800 AD. Prolonged droughts affected the southern Peru-Bolivian altiplano in the period from 1200 to 1300 AD. The most severe manifestations are those affecting the southernmost coast, as far south as Nazca, 280 miles south of Lima.

---

[207] Michael E. Moseley, *The Incas and Their Ancestors: The Archaeology of Peru*, 223.

[208] W. Quinn, *Registros del Fenómeno El Niño en el Perú*, Ifea, 17– 18, quoted in Klauer, *El Mundo Pre-Inka: Los Abismos del Cóndor*, tomo 1, p. 38.

[209] Ibid., 39.

[210] Ibid., pág. 49.

[211] Ruth Shady, 'La neolitización en los Andes Centrales y los orígenes del sedentarismo, la domesticación y la distinción social', in Ruth Shady y Carlos Leyva, editors., *La ciudad sagrada de Caral-Supe: las orígenes de la civilización andina y la formación del Estado prístino en el antiguo Perú.*

Arctic Warming: An Exception or a Recurring Phenomenon?

The North Atlantic has been a key region of warming over the decades of the 90s and early 2000s. It was affected by a positive phase of the NAO between 1900 and 1940 and a negative between 1940 and 1990.

It is quite instructive to look at the polarity and strong contrast between Arctic and Antarctica. The Arctic Ocean is a shallow ocean, with only one major ice cap over Greenland. Antarctica is a vast continent with two major and extensive ice caps and large thick ice shelves reaching into a shallow shelf sea. The North Pole is situated over a few feet of sea ice, whereas at the centre of Antarctica the South Pole is covered by 3 kilometres of ice deposits. The Arctic is surrounded by the land masses of North America and Eurasia separated by shallow straits, while the southern counterpart is extensively surrounded by 1200 miles (2000 kms) of oceans, with only the tip of South America coming near it. While Antarctica has registered very little climatic variations, the world's attention has been drawn to the dramatic warming of the Arctic, with consequent loss of sea ice.

The North Atlantic is subject to a low frequency cycle of 80 years and smaller cycles of 10–20 years. The global warming of 1980–2000 can be seen as part of a recurring cycle. The Arctic warmed up more considerably than other parts of the world, even after the peak warm years, showing delayed effects. Thus, between 2003 and 2007 there was a 50% decline in Summer sea ice. This extended to Alaska and Siberia where the permafrost started melting, and lakes and cracks formed on the Greenland ice cap.[212]

No attention was drawn, however, to the fact that the Arctic periodically undergoes dramatic cycles of warming and cooling, which do not happen in Antarctica, because the global climate patterns are amplified in the far north. Much of the heat accumulated during warming cycles is displaced by ocean currents in Pacific and Atlantic oceans toward the Arctic. The cycles of heatwaves appear around a 70–years periodicity, with the last heat wave hitting the Arctic in the 1940s.

The 2002 temperature records for Greenland's eastern coast at Godthaab were still below those of the 1940s; things were less marked on the west coast at Angmagssalik, with only the 2002

---

[212] Peter Taylor, *Chill*, 133.

yearly record surpassing those of the 1940s. These phenomena can be seen at play if one looks at the records going from 1866 to 2002. Pretty much similar results are repeated for Norway, the Jan Mayen Islands (between Norway and Greenland), or Alaska and Yukon.[213] Among the years that registered unusually warm temperatures in Greenland are 1929, 1932, 1941, 1947 and 1960, all of which had temperatures warmer than in the 80s and 90s.[214] And, of course, we have the historic record, which reminds us that a thousand years ago the Vikings' cattle could graze around their Greenland settlements.

AO and NAO were predominantly in positive phase between 1989 and 1995 and once more between 1999 to 2009. Thus Greenland's warming seems to be predominantly driven by warmer North Atlantic waters. This leads Peter Taylor to conclude: '… the Arctic warming is not unusual and is clearly affected by a long-term oscillation pattern'.[215] Antarctic specialists do not attribute the break-up of the ice shelves in the north to global warming since they can observe that there's been very little rise of temperatures in Antarctica, except on the peninsula region.

Trends in Cloud Cover and Albedo

From satellite data we know that cloud cover increased between 1960 and 1980. When it decreased after 1980 the temperatures started to rise. So-called 'albedo' is generally expressed in percentage of energy reflected by clouds or ice cover into the atmosphere. Through albedo an average 25% of the light and energy flowing at the top of the atmosphere is reflected back.

The albedo—influenced by the cloud fractional coverage—plays an important role on climate and its modification as we saw in Chapter 1. A 1% change is equivalent to 3.4 watts/sq m energy balance change, or the equivalent to the computed effect of a doubling of carbon dioxide levels. Between 1984 and 2000 there was a decrease in cloud cover, and therefore a steady decrease in albedo (an estimated 2%) with the strongest drop in the 1990s. The amplitude of the decrease was of the order of 2–3 W/m^2 to 6–7 W/m^2. Wherever we situate the change within this

---

213 Peter Taylor, *Chill*, 135–136.

214 Peter Taylor, *Chill*, 147.

215 Peter Taylor, *Chill*, 139.

bracket we are going to witness significant changes in the Earth's energetic budget.[216]

A study of E. Pallé estimates an increase of short-wave radiation reaching the Earth of the order of 6 W/m^2 between 1985 and 2000. This dwarfs the effect of CO_2, estimated at 0.8 W/m^2. In comparison, the light reaching the surface of the Earth decreased by about 2 W/m^2 in the period 2000–2004. (76, 78)

From the studies of various authors Peter Taylor concludes that:

- There was an increase of short-wave radiation (visible light) between 1983 and 2001 of the order of 0.16 W/m^{2}per year.
- This flux change was correlated to a change in cloud patterns and thickness, and in the transparency of the atmosphere.
- The change was felt mostly over the oceans, with an average 0.34 W/m^2, whereas land-based observations showed no significant rise.

Over the period 1983–2001 this change amounted to 4.5 W/m^2, compared with 0.8 W/m^2 for the computed total effect for additional CO_2 (or greenhouse gases).[217]

These are key findings for the whole question of climate change, confirmed in many places and by various researchers. Although these cloud data are not regarded as controversial, the concerns they could raise are not considered in the IPCC Summary for Policymakers, whereas they emerge in the Working Groups. In the Working Groups Peter Taylor reports there is much stronger agreement on cloud changes and radiation flux at the surface than on almost any other topic.[218]

The View from the Sun: Solar Cycles

As in the case of the effect of ocean currents the evidence of the impact of solar cycles started to grow at the time in which the models were constructed, leading to the realization that models should be changed. There is now growing evidence of a link between solar cycles and ocean cycles. The mechanism is most likely indirect through the influence of cloud patterns.

In addition to sending short-wave radiation (visible light) and UV radiation the Sun generates the 'solar wind', a plasma of high-speed electrons and protons that accelerates as it moves away from

---

[216] Peter Taylor, *Chill*, 74–76.

[217] Peter Taylor, *Chill*, 82.

[218] Peter Taylor, *Chill*, 58–59.

the Sun's surface. This energetic plasma moves along magnetic field lines to form a narrow spiralling 'skirt' that reaches out from the solar equator to the plane of the ecliptic across the whole solar system. This renders the solar system similar to an invisible bubble moving in space in which the larger planets bring into effect a mesh of variable magnetic and gravitational effects.

The Sun counts three major solar cycles, among a plethora of others:

- The Schwabe cycle (11–year average) is the famous sunspot cycle. Although its average is of 11 years, it varies between as little as 9 and as much as 13 years. The sunspots decline over 4–5 years to the point that there is either absence or very low number of sunspots for a few months.
- The Hale cycle (22–year average) is simply due to the fact that from one Schwabe cycle to another the Sun reverses its magnetic polarity. It takes 22 years in average to return to the original polarity.
- The Gleisberg cycle (varying roughly between 80 and 105 years) modulates the length of the Hale cycle and the intensity of the 11–year cycle.
- There are other periodicities of the order of ~ 400, ~ 800 and ~ 1500 years, as inferred from proxy data. We will return to these and still others in Chapter 9.[219]

Among purveyors of proxy data of solar wind activity—the plasma flux of electrons and protons—are the so-called 'cosmogenic isotopes' of beryllium and carbon created when cosmic rays bombard the atmosphere. Here we find an inversely proportional relationship. When the solar wind is strongest at the sunspot peaks the geomagnetic field rises and deflects cosmic rays, leading to a decrease in the rate of production of the cosmogenic isotopes. Research on solar activity highlighted a clear correlation between the 1983 to 1996 solar cycle and a 3% increase in cloud cover, itself correlated to the cycle of the solar wind. At the peak of sunspot activity the solar wind increases, at the minima it drops.[220]

In the 1990s scientists discovered correlations between the variable length of the sunspot cycle and the surface temperatures of the northern hemisphere over considerable time spans. The Sun's solar visible emissions in relation to the strength of its electromagnetic

---

[219] Peter Taylor, *Chill*, 162–64.
[220] Peter Taylor, *Chill*, 167.

field correlate with the so-called 'Maunder Minimum' of solar irradiance, which fell between roughly 1610 and 1780, and rose in the recovery period leading to the present 'Solar Grand Maximum'.[221]

Relating to the above is the discovery of Mike Lockwood's team at the Rutherford Laboratory in England, recording a 230% increase in the solar wind from 1900 to 2000. The long-term cyclic patterns correspond with the well-known cycles of cooling and warming such as the Little Ice Age and the Medieval Warm Period.[222] There has been a continuous rise in the solar flux after 1950 and the peaks observed after 1990 are the highest since 1700. After 2004 there has been a sharp rise of cosmic ray count, which very likely created a rise in low-level cloud cover—compared to previous cycles—causing a cooling effect. Taylor indicates that by 2009 temperatures had begun to fall from historical high points.[223]

The View from the Cosmos

The latest scientific revelations become almost staggering when it comes to the Earth's relationship with the wider cosmos, and what can be measured of it. The later decades of the twentieth century have witnessed an increase in electromagnetic radiation from a strong solar wind. This resulted in a transfer of strength from the solar and interplanetary magnetic fields to the Earth's magnetic field. On the contrary the beginning of the twenty-first century was accompanied with a long absence of sunspots and the lowest levels of geomagnetic field strength on record. Other evidence has emerged that corroborates these findings backwards. The magnetic field varies in strength over the matter of centuries, with a low point corresponding with the Little Ice Age, and a high point with the Medieval Warm Period.[224]

Based on the deposition of beryllium isotopes in proxy records researchers have recognized two lows of galactic cosmic ray activity at 1400 and 1700, the second being the strongest. These data correlate with proxies for the determination of temperature in the ocean sediments. 'Thus we can see a clear correlation between increasing solar flux, the decline in cosmic ray flux and lower beryllium-10 with rising temperatures, and the inference is that the

---

[221] Peter Taylor, *Chill*, 167.

[222] Peter Taylor, *Chill*, 166–167.

[223] Peter Taylor, *Chill*, 174.

[224] Peter Taylor, *Chill*, 180.

declining cosmic reflux leads to less cloud cover and much sunlight reaching the oceans to be stored as heat.'[225] These results are a confirmation of cloud effects pointed to earlier in this overview. Some authors advance evidence that the recent rise in solar magnetic activity (1980–2000) is the highest in 8,000 years, and it is exceptional both in amplitude and duration. But it is also likely to have plateaued at present.[226]

Researchers at the Oulu University in Finland have detected correlations between clouds and galactic cosmic rays (GCR) and a possible process that links solar variability with climate. In their discoveries low clouds are statistically related primarily to GCR and to a minor extent to ENSO (El Niño Southern Oscillation). The GCR acts through ionization of the atmosphere. These correlations have been found to be significant over the years 1990 to 2010. The variations of low cloud cover have been found to correlate with the 11–year Schwabe solar cycle. When one looks at the Arctic region the variation of low cloud cover exhibits a 'highly significant one-to-one relation with inter-annual variations in the ionization over the latitude range 20–55° and 10–70°', leading to the hypothesis that the cosmic ray induced ionization modulates cloud properties.[227]

An Alternative View to Computer Models

On the basis of all the mounting evidence we can come to the conclusion that the CO_2 forcing effect has been overestimated by a factor of 3, if not 4 or 5. Instead the main drivers of the warming have been cloud cover and oceanic effects, with their currents transferring the heat.

The *assumption* of an amplifying effect in the interplay of CO_2 and water vapour estimated at 300% is a key fixture of models. However, it is entirely theoretical, inbuilt into the models, not derived from external reality. The facts rather show the contrary. The relationship of CO_2 concentrations to radiative forcing is logarithmic. Past a saturation point, the power of CO_2 to warm up the atmosphere diminishes rapidly by a factor 10 for each additional unit of CO_2.[228]

---

[225] Peter Taylor, *Chill*, 183.

[226] Peter Taylor, *Chill*, 185.

[227] Peter Taylor, *Chill*, 186–187.

[228] Peter Taylor, *Chill*, 42.

If we look at millennia rather than just the last 150 years, as IPCC reports do—in other words if we add up the evidence of non-instrumental data to the more recent instrumental data, plus available historical records—the main drivers of warming in our time (and mostly in the period between 1980 and 2000) are put in evidence by satellite data, especially the Satellite Cloud Climatology Project. They can be recognized as:

- Past solar magnetic highs and lows that correlate with temperature variation proxies over millennia. The influence of the longer-term cycle of solar heating, the present Solar Grand Maximum. This is due to an increase of 230% of solar wind (the plasma flux of electrons and protons) from 1900 to 1990.

- An unusual combination of natural cycles of warming in the first place in northern hemisphere ocean basins and the teleconnections between these oscillations, that have peaked together, which have driven the recent decades of global warming. Ocean oscillations and periodicities and their effects on temperatures (especially in the northern hemisphere) have now been amply documented: among these ENSO (4–8 years), NAO, (~20– and ~70–year cycles), AO (60–70–year cycles), PDO (Pacific Decadal Oscillation of ~30–year period). These can amplify or contrast each other. Thus the PDO in its negative (cold) phase was in place during the 1945–78 cold period; the warm phase coincided with the period of global warming of 1980 to 2000.[229]

- Satellite data have recorded a low cloud-cover decrease between 1980 and 2000, accompanied with increased short-wave radiation to the surface of the ocean, whose total could account for all of the warming in the corresponding period. Even just a 1% drop in reflective low-level clouds can produce a warming equal to what is attributed to CO_2 through computer modelling. And models do not have a reliable cloud component that would take into account cyclic variations, even by the experts' admission. Keep in mind that the cloud cover decreased by 4% between 1983 and 2000. The increase in visible short-wave radiation is enough to drive the whole of the late twentieth-century warming. This increase in radiative effect is enough to dwarf the CO_2 effect by a factor of 4 to 6.[230]

---

[229] Peter Taylor, *Chill*, 216.
[230] Peter Taylor, *Chill*, 72–73, 88.

- The solar heating occurs in cycles of 11 years and is influenced by a little-understood solar-terrestrial mechanism that has reduced cloud cover, sufficient to create a strong warming pulse over the period 1980–2000.[231] The 11–year cycle fluctuation in the solar wind is responsible for a 3% variation in low-level reflective cloud cover. The pulse is clearly evidenced by the satellite data.[232] The two 11–year cycles formed a peak in 1980 and in 2000. The subsequent drop in global temperatures and loss of accumulated heat in the oceans is parallel to a 30% fall in solar magnetic activity from the years 1990 to 2000.[233]

The above cycles brought about the accumulation of warm waters in the oceans. This warmer water moves, for the most, toward the northern hemisphere, especially the Arctic regions. Time lags, interferences between the cycles of the oscillations and amplifying patterns resulting from these have brought about the warmer spells toward the end of the century.

Through the extended perspectives reached in this chapter, two new boundaries are reached in Peter Taylor's estimate: 'We would do well to bear in mind therefore that however much we may feel the boundaries of climatology have been delineated, there are still major cycles that lie beyond the confines of our solar system. At one time, the sun was regarded as a constant in the equations, now we know that he has cyclic variability.' And further: 'Finally, the sun itself is not an isolated system—as modern astrophysics has discovered. Its cycles and periodic activity are unlikely to be as internal to the sun or even the solar system...'[234]

The present compendium of existing research has offered us an inkling that the Earth is not a closed system, not according to the Second Law of Thermodynamics, which we saw invalidated in Chapters 1 to 5, nor according to its relationship with the solar system and what lies beyond it, which we just explored.

On one hand, in all living systems a great variety of phenomena contradict the Second Law of Thermodynamics, which predicates the conservation of energy. The experiments of von Herzeele and Hauschka on one hand, and all the work and practical applications

---

[231] Peter Taylor, *Chill*, 48.

[232] Peter Taylor, *Chill*, 53.

[233] Peter Taylor, *Chill*, 217.

[234] Peter Taylor, *Chill*, 178.

of Schauberger flatly show us that this construct is only true in mechanical systems, not in Nature. Here appears a first flaw in the ground upon which computer models predict our climate and its possible evolution.

To the challenge in this direction is added, and abundantly proved, that the Earth is not a closed system. In fact climatic variations *cannot* be understood—not only in relation to the last 150 years, but millennia backward—if we do not take into account influences affecting the Earth from the Sun and the planetary system. Want it or not, the rug has been taken from under our feet if we cling to a purely mechanistic understanding of the Earth's climate. So what conclusions are we to draw, even if tentatively, as is necessary in such a complex and encompassing field of inquiry?

The wealth of new revelations warrants a review of the science of climate change. We will turn to this in the next chapter, after which we will revisit the findings of this chapter and place them in the perspective of a new Goethean perception of Earth-solar-planetary relationships.

Chapter 8

Reviewing the Science

> *'It is no longer sufficient to discover basic laws and understand how the world works in principle. It becomes more and more important to figure out the patterns through which these principles show themselves in reality. More than just fundamental laws are operating in what actually is.'*

Otto Peitgen and Peter Richter

If we look only at the last 150 years of the instrumental data the recent warming period appears highly unusual. This is what was presented by the IPCC's Third Assessment Report of 2001. In the years that followed (2001 to 2007) this approach was heavily criticized. The IPCC itself now admits the evidence of greater variability in the past and the uncertainty around natural cycles, even around the Medieval Warm Period about 1000 years ago and the Little Ice Age of only 400 years ago.[235]

The global warming signal prior to 1950 is now regarded as predominantly natural, a rebound from the Little Ice Age of 1400–1800 AD. The dip between 1950 and 1980 is also presently considered of natural origin. And between 2000 and 2008 there was no net warming; then the validity of the models rests completely upon the pattern 1980–2000, completely abstracted from its relationship to the whole.

Cooling and Warming in Relation to Longer Cycles

When I studied ecology first, then environmental sciences I naively thought that I could acquire a view and understanding of the whole by sequentially adding the parts. I wanted to acquire an encompassing understanding of Gaia's ecology, so to speak. That meant looking at the large constituent blocks in addition to the general ecology: plant physiology, ecology and study of world ecosystems, oceanography, followed by river ecology, alpine ecology and forestry on the other to cover the most important bases, while still leaving others aside. Since the human factor must be taken into consideration in my Masters I added studies on pollution of all sorts and remediation

---

[235] Peter Taylor, *Chill*, 30.

and treatment of solid, liquid and gaseous waste. While I kept adding the parts I never felt I was building the bridges I needed, which could lead to a wider understanding of the whole. They remained disjointed parts. What models simulate is reminiscent of this approach.

With the benefit of a scientific background I have drawn in this study two very rich strands that overlap and complement each other. They considerably widen the picture that builds up an understanding of climate beyond current standards, though they still do not fill the whole picture. Remarkably, the record of interdisciplinary scientific studies and what emerges from a Goetheanistic/phenomenological approach to world ecology offer convergent results.

Before delving into this convergence in more detail, let us notice first of all that the most advanced of modern science, meaning the one that encompasses the widest nexus of relationships and correlations indicates that it is water, and not CO_2, that is the major factor in determining climate. Furthermore, a great part, if not the greatest, of the disruption of the water budget worldwide depends on factors that lie beyond human control. We live in an open system in which the solar system impacts world climate by the intermediary of solar flux and visible solar energy in cycles that we have started to understand and correlate to climate variability not just over the last one hundred and fifty years, but over millennia. The mix of instrumental and non-instrumental data (proxies) converges toward an unmistakable common factor: the balance of water and heat in the atmosphere and in the oceans. With these we can understand why the last one hundred and fifty years have witnessed periods of warming and cooling, and why even the most spectacular changes, like those in the Arctic region, are not one of a kind, but recurring phenomena. Under this broad perspective climate can actually be understood through the effect of changes in the water balance. Carbon dioxide has a much lower impact when compared to water.

Interestingly, the above has emerged in looking at a whole-system perspective that starts from the oceans, moves up to the atmosphere and expands into the solar system. In the Goethean approach, to which Schauberger has contributed the most, we have looked at human influence over the land masses: forestry, agriculture, water management and production of energy. Here too a direction has emerged with clarity and consistency. Whether we look at how forests and farmland are managed, how rivers are regulated and how energy is produced, once more it is the water cycle that is the most affected, and this time by the human being. Here we have outlined the pervasive displacement from full hydrological cycle to the half hydrological cycle. This has obvious consequences on climate patterns. However, what Schauberger has pointed to could hardly be accounted for from present data, because we would need to gather them differently; e.g., changes not in total precipitation amounts but in the patterns and distribution of rains at the local level; total precipitations for single events, geographical variations, lengths of draughts or floods, etc.

Water plays a role far more important than CO_2, and this causes effects even more insidious than climate change, because they pass unnoticed. This is why we have coined the term 'planetary depletion'. The universal shift from the full hydrological cycle to the half-cycle has brought in its wake, the sinking of the water table, the general disappearance of springs, the ecological frailty of forests and farmlands, and what we could call an overall depletion of the Earth's etheric. It is to this latter aspect that the term planetary depletion applies.

An example will illustrate what is meant with a depletion of the etheric. A plantation is not only different from a forest in its species composition. It differs in its ecological budget and behaves very differently in relation to how it replenishes its water horizon; whether water is readily available year-round in the soil or whether the trees mostly assimilate it from the rainfall; finally in relation to how water and nutrients move through the tree itself. Upon these premises, in a plantation rather than in a forest we will have negative consequences on the health of the trees and soil, and the quality of the finer etheric energies that are produced or most likely only poorly so. An evidence of this is massive mono-species tree die-back, even in largely forested areas. Plantation trees that have lost their etheric liveliness, will die back, according to Schauberger, because they simply cannot hold on to the soil as they naturally would otherwise. For one, water is no longer present in the soil horizon as it would have been, meaning year-long. Secondly the tree's cycle of water, nutrients and energies is depleted, and so is its etheric energy, rendering it far more susceptible to droughts and winds. The tree falls because it no longer has the etheric energy to withstand external challenges and hold on to the soil. This is obviously something that conventional science cannot fathom.

Planetary depletion is pervasive because it touches on forests, rivers and farmlands. In the present circumstances the cultural awareness that would lead to change is mostly absent.

Something else has emerged from the two complementary approaches. If we take into consideration the larger modern-scientific perspective, and if we account for the full range of phenomena that should be taken into account, then we are facing a level of complexity that simply defies imagination. This has profound consequences on the science of climate modelling. Let us look at this in stages.

The Limitations of Model Thinking

As we are approaching a review of climate change as presently understood we can look at the climate models' limitations when they do not include the reality of the Earth's interconnection with ocean cycles, solar cycles and even beyond. To do so we will first look at the inherent limitations even when we only look at the Earth from the perspective of physicists and modellers.

As the science of climate modelling becomes more and more complex and encompassing, so do the challenges to the integration of the whole. Each generation of models incorporates more and more data and sets of correlated relationships and yet the $CMIP_5$ (Coupled Model Intercomparison Project$_5$) group of models gives more unreliable predictions than two generations of models earlier ($CMIP_3$). Already $CMIP_5$ model ensembles 'cannot be taken as a reliable regional probability forecast' and that in relation to any aspect of the climate.[236] Keep in mind we have reached the $CMIP_6$ generation! The more evolved, and supposedly sophisticated, the analysis and integration of variables, the higher the resulting degree of uncertainty.

To be rigorously scientific a model should start from an observed baseline; it should be 'initialized' with accurate data at all points of the system and this is simply not possible with present technology. Next, as we saw in Chapter 1, comes the problem of 'tuning', a sort of adaptation of the system. Since the models' results rarely match what we observe in the climate, they have to be adjusted to

---

[236] Steven E. Koonin, *Unsettled*, 145.

the recorded data. Most often, this is much more than fine-tuning. And here the door is opened to adjusting the model to the results that are desired and compensate for its shortcomings. Stephen E. Koonin calls this 'an unavoidable dirty part of climate modeling, more engineering than science, an act of tinkering that does not merit recording in the scientific literature'.[237]

Other problems add up in a methodology that most often remains a mystery to the layman:

- The assessment reports do not derive from one given model; rather they are the average of a few dozen different models and they are compiled by the CMIP. This means averaging models that reach very different results from each other, and many of them from observations. This leads us to ask ourselves: if models mimic observations on the field, why would we need more than one?
- The problem of the range of results: just the variable of simulated global average surface temperature can differ about 3°C between models, which is three times the observed value of the twentieth century warming they are trying to replicate.
- Models don't work well with anomalies within the larger trends. An example: most models cannot replicate the strong warming registered in the years 1910 to 1940.[238]

The last example highlights another challenge. Climate change is used in such a way that it automatically means 'human-made climate change'. The 'Global Mean Sea Level', which is often highlighted as an alarming trend of human-generated climate change, has been rising since the end of the nineteenth century, well before significantly human-generated climate change. Some argue this may have started as early as the late 1700s. It now increases at a rate of 1.8 mm (0.07 inches)/year.[239]

The above is just one set of obvious limitations. When it comes to the growing reality of planet Earth as an open system in connection with Sun and solar system and far from a static equilibrium, the models bring other unpredictability, due to their reticence to accommodate for a growing body of evidence. Whereas speaking of discrepancies attributable to statistically insignificant variability would have been acceptable in the 1990s, we know twenty or more

---

237 Steven E. Koonin, *Unsettled*, 85.

238 Steven E. Koonin, *Unsettled*, 88.

239 Steven E. Koonin, *Unsettled*, 154–56.

years later that much of this is not random variability, but part of the changes due to cycles that are known and that actually account for it. And yet, even the IPCC's 2022 Sixth Assessment Report—both in its Technical Report and its Summary for Policy Makers—still speaks of 'natural climate variability'.[240]

There is no mention of cycles—other than 'projected climate-driven water cycle changes'—nor of impact of cloud cover or ocean cycles or currents, in the report, other than a reference to 'anthropogenic warming contributed to climate extremes induced by the 2015–2016 El Niño, which resulted in severe droughts'.[241]

Climate models are still predominantly physical models that ignore the living reality of cycles, pulsations, periodicities, including variations of these that go beyond Earth itself. In fact, there are various aspects of Earth ecology that the models completely ignore. Here is a non-exhaustive list:

- Models assume an 'equilibrium state' that will be altered by any increase in greenhouse gases, trapping more heat and raising the global temperature. However, this assumed equilibrium state is not found in the natural state of the planet because the Sun's energy output varies, affecting cloud cover and albedo, and because there are delayed responses of the world's ocean streams, which retain and transfer the heat. By their nature, cycles are such that they would oblige the models to consider a 'permanent lack of equilibrium'. This would question the nature of present modelling.
- The relation of warming oceans, water vapour and clouds has not found a place in the climate models. If these were incorporated we would realize that the power of CO_2 to influence climate has been largely overestimated. Under this realization it would emerge that CO_2 is not the major greenhouse gas. Rather we would find this to be water in the form of water vapour, mediated by clouds.
- The longer-term influence of solar cycles, in particular the variations in visible light which influence ocean warmth and solar magnetic flux, and likely affect cloud cover, are not taken into account by models. At present the models would not account for the great disparity between the Arctic and Antarctica.

---

[240] See https://www.ipcc.ch/report/ar6/wg2/downloads/report/IPCC_AR6_WGII_SummaryForPolicymakers.pdf
[241] See https://www.ipcc.ch/report/ar6/wg2/downloads/report/IPCC_AR6_WGII_TechnicalSummary.pdf

- The matter of time considered: the rise of ocean surface temperatures from 1950 to 2007 looks impressive in relation to the 1850–2007 period; if the patterns of the last 10,000 years were included then attention would be drawn to peaks and troughs in 1,500– and 400–year intervals and the 50–year signal would be submerged in the whole.
- The models have been hailed successful in hindcasting the major dip in temperatures between 1946 and 1978. At the time they used the assumptions of the effects of the increase of sulphur emission and the impact of aerosols, and their effect on reflecting the sunlight to replicate the period of global dimming.[242] This says little about the validity of the models since we now know that the dimming was not caused by sulphur emissions, which were confined to continental areas, and moreover regionally. All these things considered, the dip still remains a mystery and/or a glaring anomaly if our gaze is confined to present, prevailing assumptions.
- Predictive modelling has already failed in the past in matters concerning the circulation and dispersal of heavy metals, persistent toxic organic chemicals and radioactivity, in spite of massive investments in monitoring and simulation. This should give us pause to think.

Most of all models cannot possibly accommodate for the growing understanding of the complexity of cycles and their interconnections, leading Peter Taylor to conclude: 'Since the key mechanisms relating to solar energy, to cloud formation and the teleconnections of the ocean basins are not amenable to mathematical formulations, virtually none of these factors can be reliably built into computer models.'[243] More specifically, William Kininmonth in his book *Climate Change: A Natural Hazard* echoes that predictive models 'by their reliance upon linear computation they cannot model the complex, nonlinear response system such as climate where the water vapour cycle is at the heart of things'.[244]

Climate change predictions become even more questionable in the body of literature known as 'event attribution studies', a growing specialization of climate science. These combine climate modelling and historical data and most often focus on weather disasters. The IPCC's 2012 Special Report on Extreme Events shows the problems

---

242 Peter Taylor, *Chill*, 211.

243 Peter Taylor, *Chill*, 217.

244 Quoted in Peter Taylor, *Chill*, 217.

and limitations of these studies: 'Many weather and climate extremes are the result of natural climate variability … Even if there were no anthropogenic changes in climate, a wide variety of natural weather and climate extremes would still occur.' The same is echoed and amplified by the World Meteorological Organization: '… any single event, such as severe tropical cyclone [hurricane or typhoon], cannot be attributed to human-induced climate change, given the current status of scientific understanding'.[245] And yet we routinely hear of extreme weather attributed to climate change.

Problems are compounded when models are used in multidisciplinary approaches to forecasting. We are so used to hearing models' predictions in the news that we do not question their obvious limitations; e.g., how can we reliably predict deaths, future agricultural disasters or economic impacts of climate change? How can recourse to inter-disciplinary modelling hold validity when one discipline alone (climate change) already leaves so much to be desired? Below are some examples that could be multiplied ad nauseam.

Before he was director of Energy Policy at the University of Chicago, Michael Greenstone confidently claimed that climate change would claim 'an additional 85 deaths per 100,000 in 2100'.[246] Part of the reason this is hardly possible to estimate is that due to accurate weather forecasts, preventative measures and better medical care, deaths related to weather events—droughts, floods, temperature extremes, …—have been drastically lowered and estimates from 2010 to 2020 were as low as 0.16 deaths per 100,000, or 500 times fewer than Greenstone's forecast. When we refer to the research paper behind Greenstone's statements we read the authors' reservations such as in: 'our full set of estimates reveals a remarkable degree of uncertainty' and some of it 'fundamentally irresolvable'. In fact the 85 reported deaths estimate of Greenstone is the one deriving from one implausibly high emissions scenario. Under a more realistic scenario the numbers shrink from 85 to 14.[247]

Similar to the above is the prediction of future agricultural disasters: an example of this comes from the Summary for Policymakers of the IPCC's Special Report on Climate Change and Land (SRCCL). This was reported in *The New York Times* following the

---

[245] Steven E. Koonin, *Unsettled*, 99.

[246] Steven E. Koonin, *Unsettled*, 169.

[247] Steven E. Koonin, *Unsettled*, 169–71.

alarmistic headline of: 'It's already bad and it's only going to get worse .... So here is what we need to do.' And all of this in spite of the reference document speaking to the contrary and highlighting that: 'Data available since 1961 [to 2011] shows the per capita supply of vegetable oils and meat has more than doubled and the supply of food calories per capita has increased by about one third (high confidence).' It goes on to add that yields of wheat, rice and maize more than doubled during this period. Despite all of this, incomprehensibly, the SRCCL goes on to claim: 'Climate change between 1981 and 2010 has decreased the global mean yields of maize, wheat, and soybeans by 4.1, 1.8 and 4.5%, respectively, relative to preindustrial climate, even when CO_2, fertilization and agronomic adjustments are considered.'[248] This assertion rests on the spurious claim that if climate change had not taken place the increases could have been higher, hence a projected loss. Added to this is an alleged loss of quality and nutritional value due to high CO_2 concentrations and higher temperatures. Here one would want to ask: How do we really know about quality and how to measure it within the context of a highly industrialized agriculture?

Director-General of the World Health Organization, Tedros Ghebreyesus, writes in the same vein in *Foreign Affairs*: 'Climate Change is Already Killing Us.' To quote alarming numbers the article includes the deaths due to ambient and household air pollution, an example of which is indoor pollution in poor countries. The deaths are caused by cooking with wood and animal crop waste, arguably originating in poverty, not climate change.[249] Even though this is coming from such high quarters as the WHO it is likely to cast a heavy shadow of doubt on the seriousness of such claims.

A third problematic area are projected economic costs. Like the example above, but moving into even more uncertain territory, is the 2018 second volume of the Fourth National Climate Assessment (NCA2018), which generated doomsday headlines in NBC News, Fox News, *Financial Times* and *The New York Times*. Here we see coupled very uncertain climate-model projections with even more notoriously uncertain economic predictions, due to the variability of economic policies, trade, technology and governance among others and their variability among countries. The studies which were cross-referenced in the assessment, such as

---

[248] Steven E. Koonin, *Unsettled*, 174.

[249] Steven E. Koonin, *Unsettled*, 171.

IPCC scenarios suggest: 'For most economic sectors, the impact of climate change will be small relative to the impacts of other drivers (*medium evidence, high agreement*). Changes in population, age, income, technology, relative prices, lifestyle, regulation, governance, and many other aspects of social economic development will have an impact on the supply and demand of economic goods and services that is large relative to the impact of climate change.' This statement highlights just what a doubly uncertain territory the NCA2018 purports to cover with high assurance, when moreover the consensus among experts is one of minimal economic impact of rising temperatures. Volume II of NCA2018 is based on a 2017 paper published in *Science* magazine, also minimizing the effects of climate change on the economy. Under their scenario the net effect of climate change / economy (GDP) projections up to the year 2090 would amount to a delay of two years of growth.[250] Unfortunately, this is how 'truths' about climate change become enshrined in popular belief and among little-informed politicians who trust the science. If these predictions do not stand the test of truth, or not even common sense, what is the point of proclaiming them as gospel?

There is more, however, when we review the science in relation to everything that has emerged from the new frontiers of scientific explorations.

IPCC Procedural and Epistemological Biases

Part of the problem of climate models' limitations is that the IPCC resists the integration of new discoveries. One of its standard counter-arguments lies in questioning the reliability of instruments, whereas this reasoning is not taken into consideration for accepted data or even theoretical assumptions on its part. Another one lies in minimizing the current knowledge of longer-term cycles, even though the evidence is overwhelming. Theodor Landscheidt, to whom we will turn shortly, was a vocal critic of the IPCC models and their failure to incorporate solar phenomena. He accumulated the weight of evidence in this direction—new solar research—and showed how these were ignored by the IPCC. He cogently made his point already in 1983.[251]

---

[250] Steven E. Koonin, *Unsettled*, 179.

[251] Billy M. McCormac, editor, *Weather and Climate Responses to Solar Variations*, 228.

Another flaw lies in procedural assumptions. New relationships are looked at as suspect because they cannot be reproduced by modelling, whereas, when that is the case, it is legitimate to ask if the model should be questioned. Consider that the 'radiative forcing' factors are not derived from observation but from atmospheric models that assume the equilibrium state as something given. Since these are not real-world data, they should not be used as evidence to support the global warming hypothesis. The assumption of an amplifying effect in the interplay of CO_2 and water vapour estimated at 300% is a key fixture of models. However, it is entirely theoretical, inbuilt into the models not derived from external reality.[252] The modelling process cannot and should not be taken as evidence in support of a theory, even less as a criteria to dismiss confirmed evidence.

At present we are at the place of a growing fight between models and evidence, with each pulling in opposite directions. Ultimately all the foundations and assumptions built into the models derive from the point in time in which they were developed, the period of temperature rise between 1980 and 2000. If there had been no such temporary conjunction of events in larger time cycles, there would have been no phrase as 'global warming', nor arguably a Framework Convention for Climate Change (Rio Summit 1992).

While models are a bone of contention, so is the definition of scientific consensus. Here the boundaries between scientific and political are inherently crossed in the very epistemological premises. As we saw in Chapter 1, the IPCC Fifth Assessment Report (AR5) confidence levels are based on the evidence (robust, medium and limited) and the degree of scientific agreement (high, medium and low). The combined evidence and agreement results in five levels of confidence (very high, high, medium, low and very low).

Whereas, evidence alone has been the North Star of scientific proof, here we are adding the agreement between scientists as another criteria. This is the sounding call for science by consensus, rather than by solid, statistically significant evidence alone.

We have seen above an instance of medium evidence and high agreement, an egregious example of how agreement drives out evidence. Medium evidence is saying in essence that there is no ground to confirm a hypothesis; but through high agreement we can reinvest it with the mantle of truth.

---

[252] Peter Taylor, *Chill*, 212.

The level of confidence in IPCC's assertions relies heavily upon 'expert judgment', for the most, the judgment of the scientists who derived or introduced the radiative forcing into or from the computer models, or those who agree to the premises. The fight for truth therefore moves from the plan of scientific congruence to one of political agreement, from enlarging the discussion according to new data, or narrowing its focus by limiting the voices who have access to the conversation. One is left to wonder if a modern Galileo would be outnumbered by contrary consensus, and simply ignored under the new rules of the game.

The political element in IPCC scientific premises—the mixing of evidence and consensus—is a dangerous scientific precedent. Those who disagree tend to fall off the margins. And when they speak out their voices hardly count. The danger lies in politicizing the science as the examples of attribution studies dangerously show. Those who have greater access to the media may determine the narrative. We are moving away from a culturally independent science to a politically controlled majority-based opinion forming.

When we move our gaze from the IPCC to scientists all over the world, it is obvious that the science is far from settled, and also far from reaching consensus. But the weight of dissent is hardly ever given its due. A few examples will speak for many. Under the banner of the Global Climate Intelligence Group (CLINTEL) 1,600 scientists from all over the world have signed a 'No Climate Emergency Declaration' in August 2023.[253] Among these, 321 are from the US, including two US Nobel Laureates—physicists John Francis Clauser (2022 Nobel Prize in Physics) and Ivan Giaever (1973 Nobel Prize in Physics). Such a fracture on the carefully cultivated image of consensus can offer pause to wonder. All corporate media have simply ignored the news, and you, the reader, are most likely to have missed it.

Not unlike what has been advanced so far, Clauser indicates that climate models dismiss the effect of visible light reflected by cumulus clouds which, on average, cover half of the earth: the models vastly underestimate this aspect which plays a key role in regulating the Earth's temperature. CLINTEL also dismissed the narrative of global warming's impact on natural disasters like hurricanes, floods, and droughts, pointing that there is no statistical evidence

---

[253] https://clintel.org/wp-content/uploads/2021/04/WCD-A4version09202014.pdf

to support these claims. From there it went on to question climate models as policy tools.

Among the things the signatories are claiming: 'Climate science should be less political, while climate policies should be more scientific.' The seminar on climate models that Clauser was to present to the IMF on July 25, 2023 was summarily cancelled.[254] Though technically only 'postponed', the speech hasn't been rescheduled to this day.

A poll of the Heartland Institute seems to confirm that consensus on climate change is far from the claimed 97% or more.[255] The survey was done on 'professionals and academics who held at least a bachelor's degree in the fields of meteorology, climatology, physics, geology, and hydrology'. Only 59% of the people interviewed expected climate change to pose a significant threat to humanity's future. Here too, the public at large is not likely to have heard about this breach in the consensus façade.

We add a last example, which will be revisited in the next chapter in which we will turn more specifically to solar system and planetary influences. A special issue of the scientific journal, *Pattern Recognition in Physics*, titled 'General Conclusions Regarding the Planetary-Solar-Terrestrial Interaction', gathered the work of nineteen scientists in the area of climate.[256] They reached the unanimous conclusion from a variety of angles that: 'the forcing function [on climate] originates from gravitational and inertial effects on the Sun from the planets and their satellites'.[257] And further, forecasting the future on the basis of the aforementioned conclusions, they added: 'Obviously we are on our way into a new grand solar minimum [cold period]. This sheds serious doubts on the issue of a continued, even accelerated, warming as proposed by the IPCC

---

[254] https://www.newsweek.com/nobel-prize-winner-who-doesnt-believe-climate-crisis-has-speech-canceled-1815020

[255] https://heartland.org/opinion/97–consensus-on-climate-change-survey-shows-only-59–of-scientists-expect-significant-harm/

[256] General Conclusions regarding the Planetary-Solar-Terrestrial Interaction, in *Pattern Recognition in Physics* journal of 2013, 1, 205–206. The scientists in questions were Mörner, N.-A., Tattersall, R., Solheim, J.-E., Charvatova, I., Scafetta, N., Jelbring, H., Wilson, J.R., Salvador, R., Willson, R.C., Hejda, P., Soon, W., Velasco Herrera, V.M., Humlum, O., Archibald, D., Yndestad, H., Easterbrook, D.J., Casey, J., Gregori, G. and Henriksson, G.

[257] http://dx.doi.org/10.5194/prp-1–205–2013.

project.'[258] Soon after this issue the publisher closed down the *Pattern Recognition in Physics* scientific journal.

All of the evidence accumulated leaves us wondering whether the science of climate change comes from a healthy scientific engagement and debate. In reality it seems that earlier hypotheses and the preservation of previous assumptions puts a limit to the emergence of new evidence. Thus, Peter Taylor points to the fact that very little research goes to elucidate the influence of PDO (Pacific Decadal Oscillation)—which shows a clear correlation to the major downturn of global temperatures—or the topics of ocean heat storage, solar-cloud relations, atmospheric electric current, all of which have an important role in atmospheric ionization and cloud formation.[259]

We have showed in this chapter the numerous breaches to the façade of a settled science. It is high time to return to the Goethean approach upon which this work primarily rests. Now that solar and cosmic influences have been ascertained can we see larger patterns at play? Can we even see purely phenomenological ways to predict climate in the near future? For this we turn to the groundbreaking work of Theodor Landscheidt and others who follow in his footsteps. Landscheidt was an independent researcher who based his predictions on the basis of the Sun and the planets' movements and the mathematics of harmonic cycles. He accurately predicted a remarkable number of climatic events—even when largely derided—and lent weight to his criticism of IPCC's failure to incorporate solar phenomena in its models.

In his modern effort Landscheidt was reconnecting with the scientific views of Kepler. It is well-known that Kepler, apart from being an astronomer and an innovative exponent of modern science, was also an astrologer. Kepler's work extended into meteorology, mathematics, philosophy, theology to name but a few. Likewise, Landscheidt reconnected a great variety of fields of inquiry no longer conventionally linked to each other.

---

[258] http://www.pattern-recogn-phys.net/1/205/2013/prp-1-205-2013.pdf
[259] Peter Taylor, *Chill*, 208.

Chapter 9

A Holistic View of Climate: Forecasting the Future

'It is not our sensory and perceptual activity that forces Nature into a strait-jacket of mathematics, it is Nature, which, in the process of our evolutionary development, has impressed mathematics into our reason as a real, existing structure, inherent to herself.'

Gert Eilenberger

In this chapter we will explore how a new understanding of solar and planetary influences is such that it can account for much of the climate effects we have witnessed over the last decades. It may be quite surprising that we can actually predict future trends based on a better understanding of phenomena to which we have easy access and abundant data, such as basic solar and planetary cycles. In other words we are once more in the realm of observation. There is no need to extrapolate, create hypotheses and derive models that fit with great difficulty only a very circumscribed view of reality. But before getting there we need to grasp some basics about harmonics, golden ratio, and fractals.

After Kepler's and Galileo's death, astronomy had come to a crossroads. There was one route that suggested the study of a living system via harmonic and musical laws, opened up by Kepler, while the other path, followed by Galileo introduced the study of celestial mechanics. The second path was chosen by Newton and has since dominated the field.

We will now turn to a new way of understanding solar and planetary influences over climate. This approach will take two stages. In the first we want to move from theory-driven computer models of the IPCC type to new models that recognize solar and planetary influences empirically and treat them according to harmonic principles. This step places us in the direction of a better match with observed climate data and their oscillations over decades, centuries, even millennia. However, this movement alone, as we will see, is not sufficient if it's not accompanied with a questioning of the basic assumptions of modern astronomy. To move in this direction we will return to the broader underlying question of the integration of polarities, which emerged in Part I of this work. Chief among these are those of gravity and levity, and of technological versus planetary motion.

Some Basics about Harmonics

The work of Michel and Francoise Gauquelin and of John Addey, has brought astrology into the field and methodology of modern research. Astrology has come into modern times, by and large as a carry-over holistic discipline resting on a basic understanding of archetypes such as planets, zodiac constellations, conjunctions, squares, transits, etc. Safe for the modern anthroposophical approach to astrosophy and some other attempts, it had kept itself aloof from the modern-scientific method. John Addey, among others, predicated a synthesis of the old, esoteric approach to astrology with the modern-scientific method that uses analysis, and in this instance the treatment of data through the statistical method. He asserts:

> To reject the most important benefit to come out of the scientific revolution can only lead to obscurantism and superstition, and would be as disastrous in the long run as the earlier rejection of the spiritual basis of knowledge. The methods of making observations, as such, are neutral; it has been the viewpoint of those who have misinterpreted the results of observation which has caused all the trouble [that he sees in the misunderstandings about modern astrology]. The same sort of observations can, and should, be the starting point for spiritually orientated thought. What is now needed is a synthesis of what was best in both movements.[260]

Addey took on the vast database of Michel and Francoise Gauquelin and detected the influences of planets on the birth charts of successful individuals in a great variety of specialized fields. He showed clear and highly significant statistical correlations.

The vastly inquisitive mind of Theodor Landscheidt extended John Addey's field of inquiry to natural phenomena of climate and biology. For this reason some have called him a 'cosmo-biologist', though this term may be limiting since he also interested himself in astrology's relationships to a great variety of social phenomena.

Most of the relationships between celestial bodies—in our instance primarily Sun and planets—and earthly phenomena are

---

[260] John Addey, *Harmonics in Astrology: An Introductory Textbook to The New Understanding of an Old Science*, Introduction.

subjected to oscillations and rhythmic periodicities, which take the form of the sinusoid, or 'sinus wave'. The most well-known examples are those of the swinging of the pendulum, the sound emitted by a tuning fork or the motion of light. The pendulum renders this wave graphic. Let the pendulum swing from an extreme and it will describe an arc before passing through the centre—where it is found when at rest—then it will move to the same extension but in the opposite direction to the original. From there it will return to the centre and reach the original extreme. In so doing the pendulum has described a sinus wave. Another example is the movement of a planet around the Sun in its elliptic orbit: moving from perihelion (the point closest to the Sun) to aphelion (the point farthest to the Sun) and back to perihelion.

The sinus wave is described by its frequency, amplitude and phase, and its subharmonics, with which we can then understand the relationships we will see develop between planets and Sun, solar cycles and climate cycles on Earth.

Here music serves as the ground from which to build the basic understanding. The building blocks of music are not the tones, but the tone-to-tone distances, or pitch distances, also called intervals. Melody is formed by a succession of intervals that are, acoustically speaking, either consonant or dissonant. Pythagoras discovered that the successive ratios of first six integers give rise to consonant intervals.

The fundamental note is obtained when a string of a musical instrument is plucked and thus vibrates over its whole length. The frequency is seen as an x/y value, such as in the relation of one segment to another, or to the whole. If the string is plucked exactly halfway along its length, the two halves vibrate separately forming two waves. This is how the first harmonic is obtained, which sounds an octave higher.

In the first subdivision, the entire length (fundamental) relates to 1/2 the length as the note relates to its octave (example C–C^1). Below are the other harmonics with their musical names:

- 2/3: the entire length to 2/3rd of the length yields the fifth (example: C–G);
- 3/4: yields the fourth (example: C–F);
- 4/5: yields the major third (example: C–E);
- 5/6: yields the minor third (example: E–G);
- 3/5: yields the major sixth (example: C–A);
- 5/8: yields the minor sixth. (example C–A♭).

Beyond the octave there are no other intervals than 3rd, 4th, 5th, 6th. The most complete consonance within the octave is the major perfect chord: C–E–G = 4 : 5 : 6 that unites the major third (C–E) and the fifth (C–G) to the fundamental note.

Harmonics are most of all known in musical language. However, for other than musical purposes it seems better to adopt the practice of equating the number of the harmonic with the number into which the whole length is divided. If two waves fit exactly into a given period, then they may be said to represent the second harmonic of that period (Figure 22). Three waves, exactly completed in the period, represent the third harmonic, four waves yield the fourth harmonic and so forth. The higher the value of the harmonic the shorter is the wave.

Below is listed the correspondence between astronomical harmonics and musical harmonics:

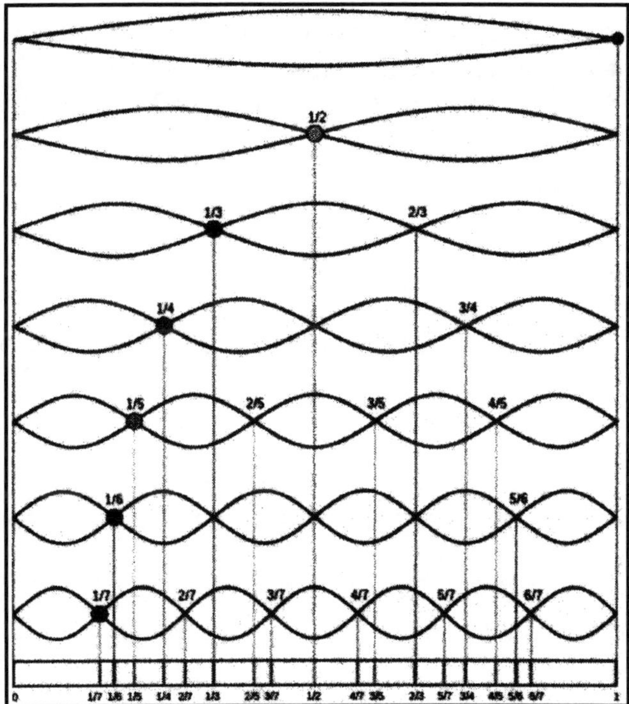

Figure 22: Sinus waves and harmonics

From the top to bottom of the diagram:

Astronomical terminology	Musical terminology
½ Second harmonic	Octave
1/3 Third harmonic	Fifth (2/3) e. g., C–G
¼ Fourth Harmonic	Fourth (¾) e. g., C–F
1/5 Fifth harmonic	Major third (4/5) e. g., C–E
1/6 Sixth harmonic	Minor third (5/6) E–G

If the length of the wave determines its frequency, the distance between extremes of the wave is its amplitude. Another way to see this is to measure the distance up and down in relation to the mean value. If the mean value is 100, and the wave rises to 120 at its highest and falls to 80 at its lowest, its amplitude will be 20%, equivalent to 120%–100% or 100%–80%, giving us the percentage of rise and fall in relation to the mean.

The last parameter of a sinus wave is its phase. In Figure 24 four waves are shown. They all have the same frequency and amplitude, but are phased differently. They have a different sequence of ascending and descending nodes, peaks and troughs (see Figure 23).

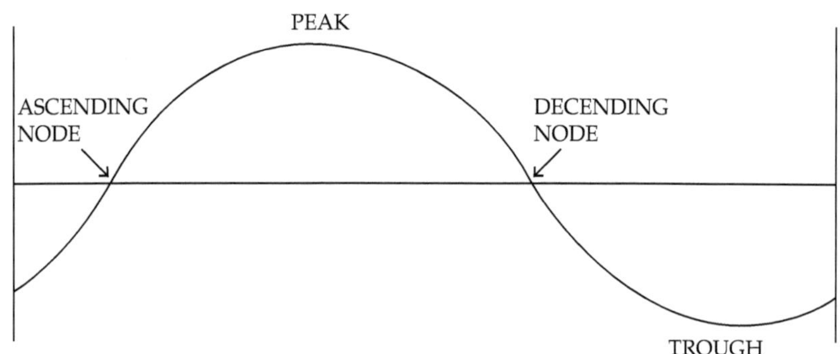

Figure 23: Parts of a sinus wave

Wave a is phased so that the ascending node comes at the beginning of the period, the peak comes one-quarter of the way along and the descending node comes in the middle. In wave b all this is exactly reversed; the peak comes three-quarters of the way along. In waves c and d the peak comes respectively at ½ or 3/8 of the way. Relating the wave to a 360° circle, wave a has a phase angle of 90° or ¼ of the circle (distance to the peak), b has a phase angle of 270°, c of 180° and d of 135°.

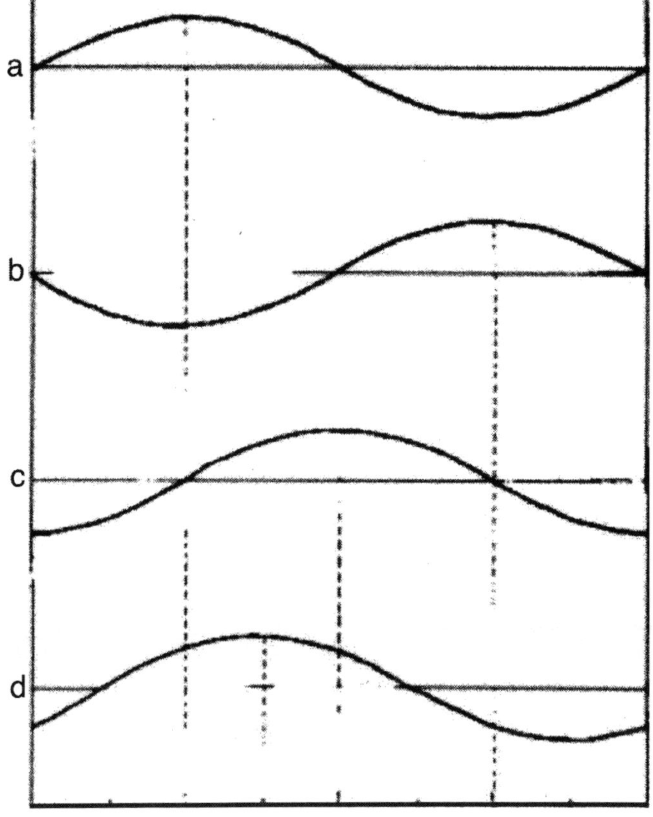

Figure 24: Sinus waves and phases

To arrive at the concept of subharmonics we can return to Figure 22, ranging from the fundamental note to the second and all the way to seventh subharmonics (astronomical terms). All of them are in phase, meaning they all start and end with the ascending nodes. In this example the 4th is a subharmonics of the 2nd, the 6th a subharmonic of the 3rd. If we extend this example the 12th harmonic is a subharmonic of the 4th, the 3rd, the 2nd because two, three and four will all divide into twelve.

At times in this chapter we will find that we are not dealing merely with a single wave form but with a complex or combination of wave forms. That is to say that we shall have to combine a given harmonic with some of its subharmonics. If we wish to combine three harmonics, we can do so in the manner shown in Figure 25. Note that in this example the three sinus waves are not in phase.

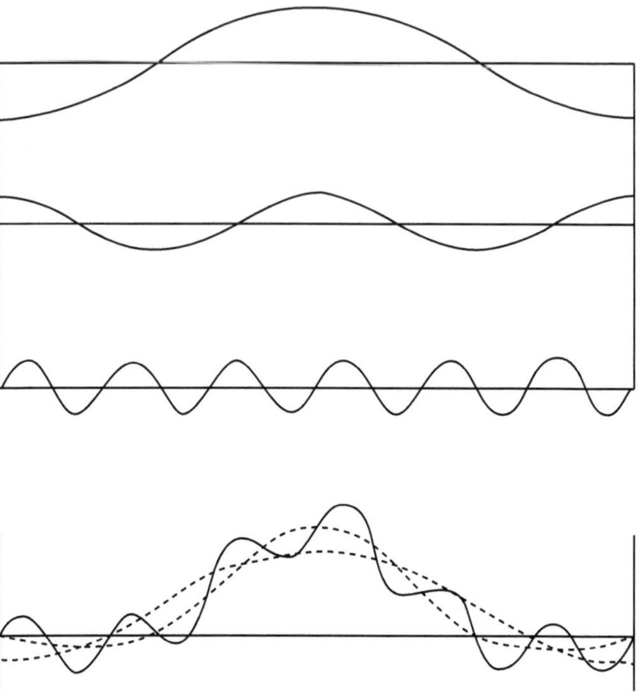

Figure 25: Building a composite wave

To continue our exploration we will now look at another very common phenomenon underlying form in Nature. It was introduced in the work of Lawrence Edwards and was also known to Viktor Schauberger.

Golden Ratio and Fractals,

The 'golden ratio', closely associated with all living systems, is found hidden in the patterns of the spiralling of the galaxy, in the form of a nautilus or many other shells, in the pattern of the sunflower head, or that of pine cones, in the human being seen as a pentagram (rendered famous by Leonardo da Vinci's drawing), to mention but a few. The ratio is derived from a unique relationship between any a and b segments, as seen in the figure below.

In the instance below $a/b = (a+b)/a = \phi$ (Phi) $= 1.618033...$ where ϕ is an irrational number. In this equation a is called the 'major', b the 'minor'. To obtain the major you multiply the fundamental segment by $0.618...$, to obtain the minor by $0.3819 ....$

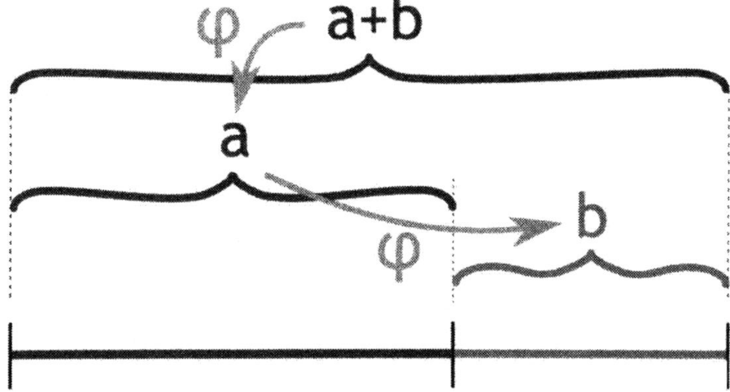

Figure 26: Golden ratio proportion

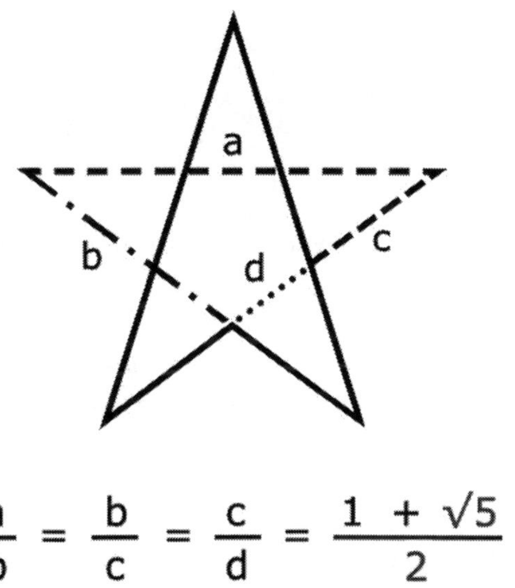

$$\frac{a}{b} = \frac{b}{c} = \frac{c}{d} = \frac{1 + \sqrt{5}}{2}$$

Figure 27: Proportions of the golden section in a pentagram

The golden section is found in fivefold symmetry, which will appear in some of the examples further on. It is present in viruses, flowers and fruits, animal organisms like starfish or sea urchin to name a few; not so, however, in the realm of crystals. Let us look at a pentagram as a basis of natural forms. This form contains a wealth of golden ratio relationships, such as those shown in Figure 27.

The pentagram can be constructed geometrically from the pentagon as shown below.

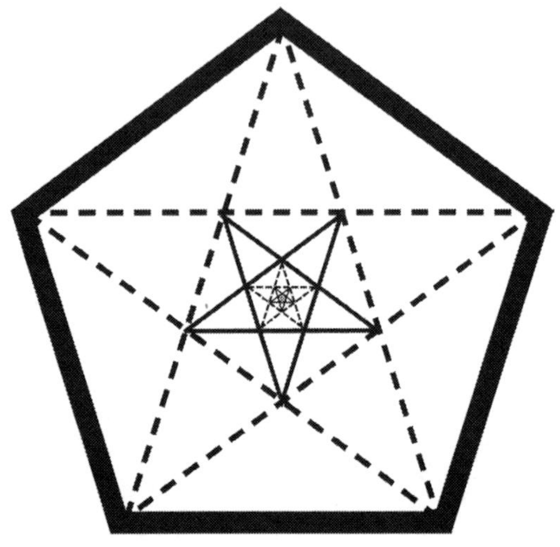

Figure 28: Pentagrams and pentagons

The initial pentagon, outlined with a thick line, served to form the inner pentagram (dashed line) that unites the corners of the pentagon. Within the second pentagram from the outside in we can recognize a new pentagon (dashed line). Within the latter we can form the medium (solid line) pentagram and the derived, pentagon (solid line), in a process that has no ending. We can in fact faintly divine two other sets of figures within.

We are hitting here upon another building block of natural forms, which has been called a fractal by the mathematician Benoit Mandelbrot. A fractal is based on the principle of self-similarity at different scales, as we have seen in the example above. In the figure above we can magnify or reduce, or iterate the process by any given factor, and still obtain the original object. A good analogy is that of Russian dolls nestled inside each other, theoretically ad infinitum. In many natural forms an arbitrary piece of the system contains all the essential structures of the whole. An example is visible in the basic structure of a fern frond. The blade structure of the whole is reproduced first in the 'pinnas' (in the order of 20–30 for each blade) arranged symmetrically around the petiole/midrib, then in the 'pinnules' (10–20 for each pinna) also arranged symmetrically within a pinna.

The above fivefold symmetry, as we will see shortly, is clearly detectable in the manifestations of solar activity. It has led Landscheidt to dispute the view of the Sun as an inert physical mass. This contention was disputed as well by Georg Blattmann in Chapter 3.

The Mandelbrot Set

If the matter of fractals is approached mathematically we arrive at the mathematics of the 'Julia sets'. In the motifs of the fractals derived from these sets, there is one constant element: the so-called 'Mandelbrot set' which appears again and again in different sizes, and marks the boundary between order and chaos. It's like a basic building block of highly organized and diversified systems (sets). To obtain a Mandelbrot, the mathematician uses a mathematical feedback cycle in which he starts from a number z_0, takes its square and adds a constant c to get z_1. He then repeats the operations from z_1 to obtain z_2, z_3, z_4,.... z_n. For each value of c a new Julia set develops. A proof that Julia sets are not abstract constructs is their appearance in Nature in the patterns of crop circles: a triple Julia Set appeared at Windmill Hill, UK in 1996.[261] Many other fractal forms, like one reminiscent of a sunflower head, have adorned the fields since.

The Mandelbrot set, at the centre of the fractal structure, captures the properties of harmonics we have outlined above. When the set is subdivided along the vertical axis, according to key segments (Figure 29) one recognizes in it all the consonant intervals of musical harmonics:

½: octave as in (TB–CH) : (LC–TB);

2/3: the fifth as in (TB–CH) : (ZN–TB);

¾: the fourth as in (ZN–TB) : (LC–TB);

4/5: the major third as in (LC–TB) : (LC–CH);

5/6: the minor third as in (ZN–CH) : (LC–CH);

3/5: the major sixth as in (ZN–TB) : (ZN–CH);

5/8: the minor sixth as in (ZN–CH) : (ZN–EA);

And the major perfect chord (4 : 5 : 6) as in (ZN–HC) : (ZN–CH) : (LC–CH).

With the above premises established the discoveries of Theodor Landscheidt will be more understandable. Born in 1927 in Bremen,

---

[261] Freddy Silva: *Secrets in the Field: The Science and Mysticism of Crop Circles*, 78.

Theodor Landscheidt studied philosophy, law and natural sciences. Apart from being a High Court German judge most of his life he was the director of the Schroeter Institute for Research in Cycles of Solar Activity founded in 1983. His research was published from 1986 to 2000, including his seminal book *Sun-Earth-Man: a Mesh of Cosmic Oscillations: How Planets Regulate Solar Eruptions, Geomagnetic Storms, Conditions of Life, and Economic Cycles*. He died in 2004.

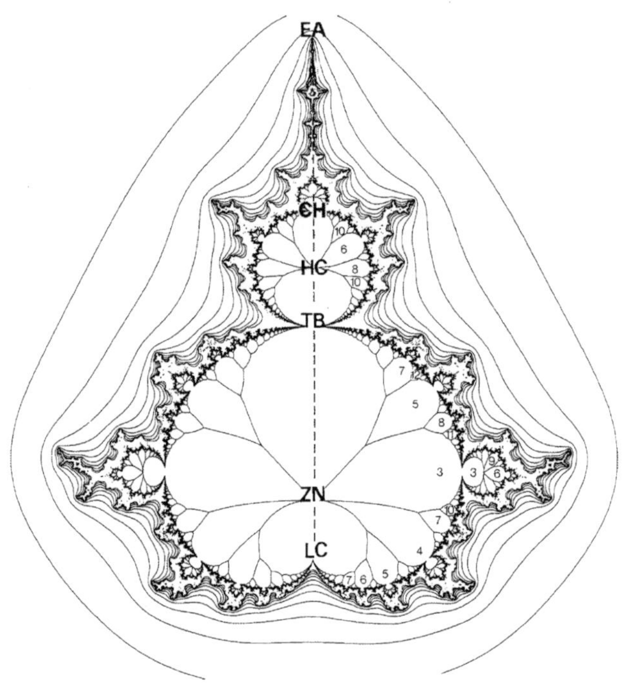

Figure 29: The Mandelbrot set

Sun and Planetary Motions

As Landscheidt predicates, we can divide the role of the planets on a functional basis between 'tidal planets' and 'giant planets'. On one hand are the tidal planets: Mercury, Venus, Earth and Jupiter, with a role similar to that of the Moon in relation to the Earth. On the other hand we have the giant planets from Jupiter to Saturn, Uranus and Neptune, in fact all the distant planets excluding Pluto, given its much smaller size. Jupiter acts as a link between the two groups. It is mostly the second set of planets, and Jupiter in particular, that will call our attention in relation to Sun cycles and solar activity.

The shortest Sun cycle, the sunspot cycle (Schwabe) of about 11 years average switches from positive to negative values between one cycle to the next. Thus, the whole cycle is really completed with 2 Schwabe cycles forming 1 Hale cycle. The Gleisberg cycle of an average of 80–90 years modulates the intensity (amplitude) of the 11–year average sunspot cycle, which is not constant but varies from 9 to 13 years.

Among the giant planets, which regulate or modulate essential features of the Sun's activity, Jupiter is what Landscheidt calls 'the weighty centre of the world of planets'. The planet plays a major role because its mass equals 71% of the total mass of the planets, and 61% of the total angular momentum of the solar system, versus less than 1% for the Sun.[262]

Jupiter also plays a key role in astronomical configurations that have an impact on the Sun's rotation, the frequency and incidence of major solar eruptions, geomagnetic storms, variations in ozone concentration among others.[263] This Jupiterian influence can be detected in long-term cycles, subdivided through their harmonics in smaller cycles. As we will see, it's in great part based on these Jupiter-related cycles that it is possible to make predictions of the Sun's activity and its impact on atmosphere and climate.

To speak about the 'Sun's motion' within the solar system is to seemingly contradict an immobility that we take for granted, and yet it is something that was already known to Newton three centuries ago, when he introduced the notion of the solar system's 'centre of mass' (CM). The centre of mass is the barycentre of the solar system. At this place it is as if the whole mass of the solar system were concentrated; and the Sun moves in relation to this centre. When Jupiter is in opposition to Saturn, Uranus and Neptune, CS (centre of the Sun) and CM almost coincide. When all four planets are nearly conjunct CS and CM are at their greatest distance.[264]

---

[262] Theodor Landscheidt, *Sun-Earth-Man: a Mesh of Cosmic Oscillations: How planets regulate solar eruptions, geomagnetic storms, conditions of life, and economic cycles*, 53.

[263] Theodor Landscheidt, *Sun-Earth-Man*, 7–8, 13.

[264] Theodor Landscheidt, *Sun-Earth-Man*, 14, 28.

When Jupiter is stationary near CM, Landscheidt noticed periods of major unstable events.[265] Usually the Sun passes through CM fairly quickly; but if CM runs almost parallel with the Sun's surface, then major solar instability events occur. These take place when CM remains within the range of 0.9–1–2 solar radii for 2.5 to 8 years. Such were the years 1789–1793, 1823–1828, 1867–70, 1933–37, 1968–72, 2002–08.[266] These solar events significantly affected the Earth's magnetic field.

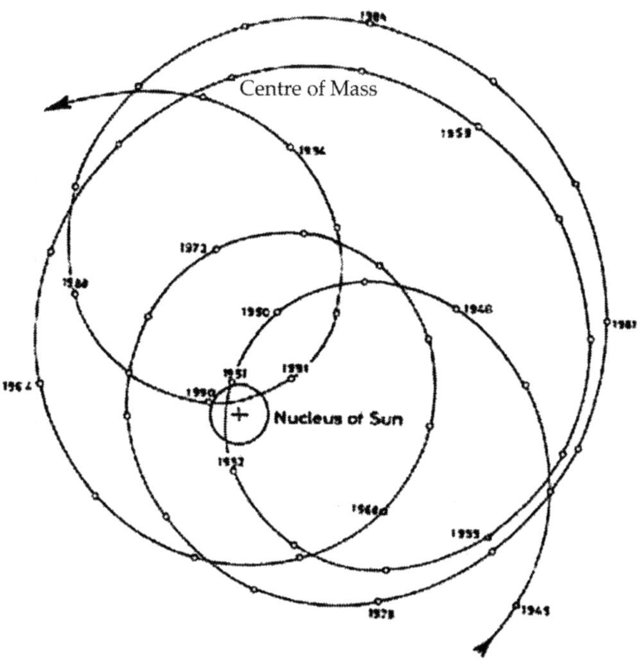

Figure 30: Motion of the Sun around the Centre of Mass

When in 1951 (see Figure 30) the Sun's centre (CS) was very near CM and then moved away from it, the sunspot cycle of 80 years reached a maximum. Approaching or moving away from such phases the orbital angular momentum of the Sun varies from

---

265 Theodor Landscheidt, *Solar Activity: A Dominant Factor in Climate Dynamics*, at https://plasmaresources.com/ozwx/landscheidt/pdf/SolarActivity_A_DominantFactorInClimateDynamics.pdf
266 Theodor Landscheidt, *Sun-Earth-Man*, 16.

-0.1 to + 4.3 x 10^{47} g/cm^2/s^{-1}or in reverse [-0.1 to -4.2 x 10^{47} g/cm^2/s^{-1}], a variation of more than 40 times in one way or the other.[267]

Central to a closer understanding of the Sun's motion around CM is the notion of 'impulse of torque'. Torque, used in the case of a rotating body, is defined as the rate of change of angular momentum, the force of a rotating object. The intensity of IOT is measured by the change in angular momentum affected by the impulse. Strong impulses of the torque take place when the Sun's centre, CM and Jupiter are inline [JU–CM–CS events], when the Sun changes type of motion from approaching toward or receding from CM. At those times new patterns can be seen at the Sun's surface.

There are two types of JU–CM–CS events taking place during heliocentric conjunctions of Jupiter and CM, that initiate strong changes in the Sun's angular momentum:

- JU–CM–CSc with sharp increase in orbital momentum and motion of the Sun away from CM. In the terms used so far in this book this corresponds to 'planetary motion' and quite fittingly so.
- JU–CM–CSg decrease in orbital momentum and motion of the Sun toward CM due to prevailing 'gravitation'. This corresponds to motion determined by gravity.

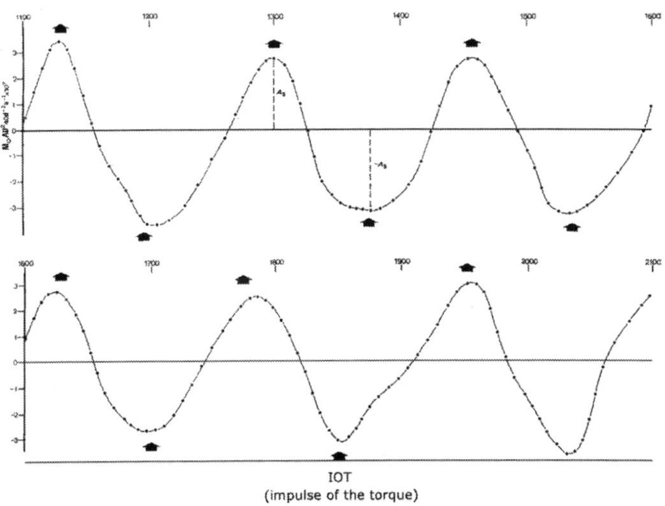

Figure 31: Time series of consecutive impulses of the torque

---

[267] Theodor Landscheidt, *Solar Activity*, 10, available at https://plasmare-sources.com/ozwx/landscheidt/pdf/SolarActivity_A_DominantFac-torInClimateDynamics.pdf

A wave pattern of the Sun's 'impulse of torque' (IOT) in relation to its movements around CM emerges. Figure 31 shows the correlation of extremes of the wave formed by consecutive IOTs with the maximum of secular sunspot cycles. For the years 1100 to 2100 a frequency (wave) of 166 years can be detected from an extremum (positive or negative) to the next. Each extremum is correlated to a maximum in the secular sunspot cycle (indicated by fat arrows in the graph). Minima in sunspot activity correspond to 0 values in the graph. The mean value of the length of the secular torque comes to 83 years—the average length of a Gleisberg cycle.[268]

In Figure 32 we see how the Hale cycle (average of 11 years) varies between 10 and 13 in the period between 350 and 1100 AD. The Gleisberg cycle modulates the Schwabe cycle, visible in the return of the wave from one phase to the next (e.g., from one ascending node to the next). Minima of this value correspond to strong sunspot maxima and vice versa. The data set are in phase.

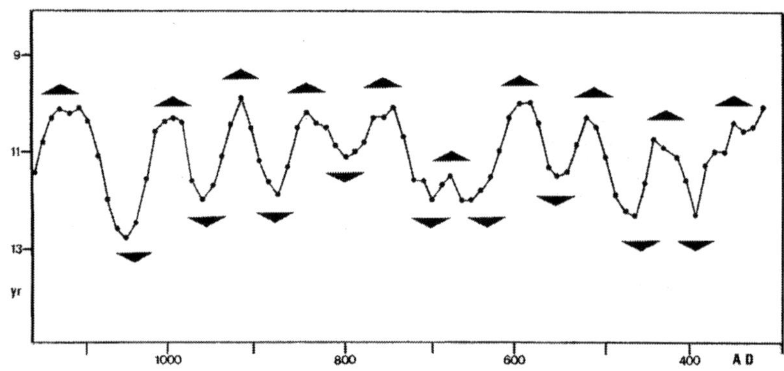

Figure 32: Secular sunspot cycles from 300 to 1100 AD.

Big Hands, Big Fingers

The pattern of Figure 33 describes the 9–year running variance of the Sun's orbital angular momentum (vertical axis) from 730 to 1075 AD. They reflect the dynamics of the Sun's motion around the centre of mass. The shape gives a repeating fivefold symmetry, which has been dubbed 'big hand' and its subdivisions 'big fingers'. This pattern has great influence on solar-terrestrial relations. A big finger cycle lasts on average 35.76 years, the big hand 178.8 years or five times as much.

---

[268] Theodor Landscheidt, *Sun-Earth-Man*, 45–46.

A cycle of 181 years exists in the sunspot data, remarkably close to the value of a big hand. And, it has been noted that the big hand cycle is almost twice as long as the Gleisberg cycle, which modulates the intensity of the 11–year sunspot cycle.[269]

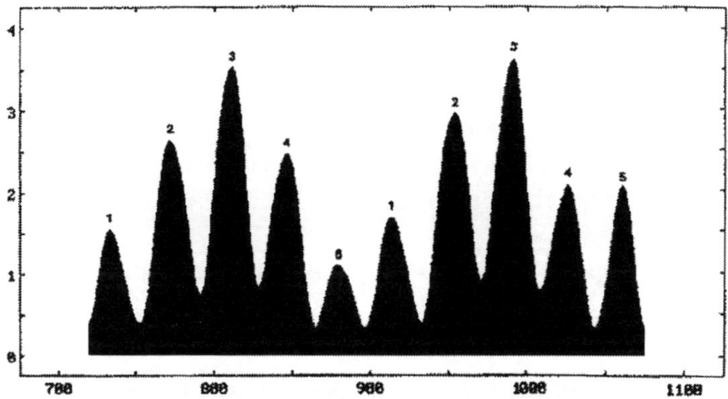

Figure 33: Cycles of big fingers and big hands

Scientists conceive that the Sun is a body of 'dead' matter. As such it should not display fivefold symmetry. Twofold, threefold, fourfold, or sixfold symmetry exist in crystals, but not fivefold symmetry reserved to the realm of the living. This unexpected extension of the domain of fivefold symmetry to the realm of 'dead' matter is all the more important since it also involves the planets, in this case most of all Jupiter which influences the angular momentum.

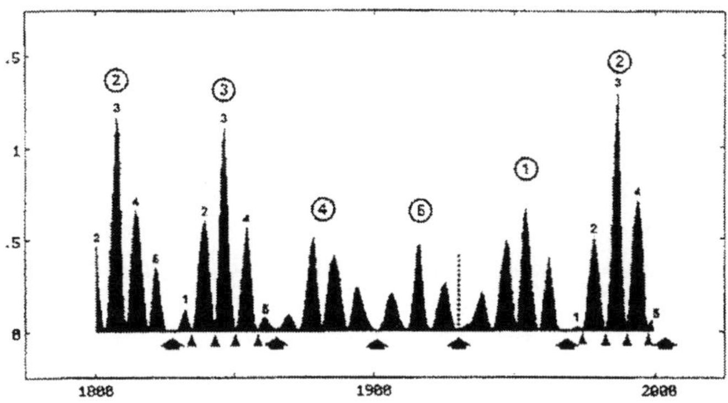

Figure 34: Small hands and small fingers

---

[269] Theodor Landscheidt, *Solar Activity*, 11, available at https://plasmare-sources.com/ozwx/landscheidt/pdf/SolarActivity_A_DominantFac-torInClimateDynamics.pdf, 11.

With fivefold symmetry also appears the fractal pattern. At a smaller scale the oscillations of the Sun's orbital angular momentum are enhanced when the 3–year running variance of the Sun's orbital angular momentum is pictured, rather than a 9–year running variance as above. Here we have not just 'big hands' with big fingers, but also 'small hands' and 'small fingers'. A big finger corresponds to a small hand when we expand the scale. The circled numbers 2, 3, 4, 5, 1, 2 correspond to tips of big fingers of the previous graph. The section from 2 to 5 (circled numbers from left to right) encloses an almost complete big hand with only big finger 1 missing.[270]

The big fingers of the big hands contain small hands with small fingers (the small, regular 1 to 5 numbers). Big arrows and small triangles designate the start of big and small fingers respectively. Whereas the previous image shows 10 peaks over 400 years, here we have some 22–23 peaks over 200 years. The vertical dotted line marks the initial phase (1933) of a big hand. Note that this nodal point coincided with the establishment of Stalin's and Hitler's dictatorship, not to mention the epochal onset of the reappearance of Christ in the etheric, according to Steiner. It will return in other examples below.

Solar Cycles

The planets, and mostly Jupiter, influence solar activity. These connections, once discovered, were tested with long-range predictions of solar activity and related terrestrial events, then announced and often checked and confirmed by astronomical and geophysical institutes. In many instances relationships around Jupiter's conjunctions with CM form a critically determining factor.

As we have seen, the shortest Sun cycle is the sunspot cycle (Schwabe) of 11 years average, which switches from positive to negative values between one cycle to the next. Sunspots have been shown to be important in relation to the Sun's effects on irradiation. More important than the number of sunspots, however, is the amount and strength of solar eruptions, or flares.

Here the researchers look at the magnetic energy present in or above sunspot areas. This is a potential which is released during solar eruptions and which influences earthly geomagnetic fields.

---

[270] Theodor Landscheidt, *The Golden Section: A Cosmic Principle*, 2.

Energetic solar eruptions, also called flares, are a sharper criterion of immediate solar-terrestrial relations than sunspots. Large flares, indicates Landscheidt, increase UV and X-ray radiations; the latter can only be detected by satellite instruments. Those recorded in the period 1970 to 1982 show that the most energetic solar eruptions come close to sunspot minima.[271]

Solar Eruptions and Solar Wind

Related to the solar eruptions is the solar wind, which increases at times of solar eruptions. The strength of the solar wind is the main factor influencing the amount of incoming cosmic rays which modulate cloud coverage as we saw in the previous chapter. At times of lowest solar eruptions the cosmic rays that reach the Earth increase. The strong influx of cosmic rays is reflected in the forming of a larger cloud cover; conversely a decrease in cloud cover takes place when cosmic radiation is lower.

Solar wind also modulates the strength of the geomagnetic field. There is a correlation between the velocity of the solar wind and the so-called 'Kp index' of geomagnetic activity. In fact geomagnetic storms occur in synchronicity with solar eruptions. They have marked consequences on Earth, where they can increase electric potentials in the atmosphere and destroy vital communication links.

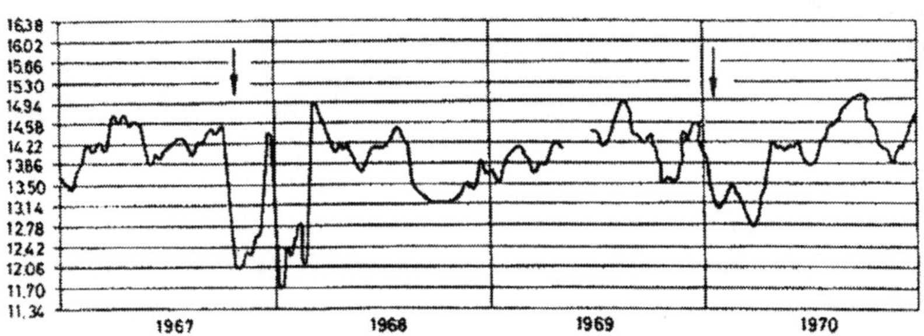

Figure 35: Evolution of the Sun's orbital angular momentum

In Figure 35 the vertical axis measures the Sun's orbital angular momentum.[272] At the starting points of small fingers cycles (SFS)

---

[271] Theodor Landscheidt, *Sun-Earth-Man*, 39–41.

[272] Theodor Landscheidt, *Solar Activity*, at https://plasmaresources.com/ozwx/ landscheidt/pdf/SolarActivity_A_DominantFactorInClimateDynamics.pdf

the Sun's orbital angular momentum reaches extrema. In the graph, two successive phases at the end of 1967 and one at the beginning of 1970 (indicated by arrows) with drastic changes in rotational velocity were correlated with heliocentric conjunctions of Jupiter with CM. At these points there take place sudden changes in the Sun's rotational velocity, related to the Sun's activity. It appears energetic solar eruptions are much more frequent around SFSs.

Based on the above knowledge, and even though these events take place at quite irregular intervals, Landscheidt's forecasts of energetic solar eruptions, covering the years 1979 to 1981 were published one year in advance. They predicted the incidence of energetic solar X-ray bursts [flares] and proton events during certain times of the year; 27 of 29 events corresponded to these areas. The same was repeated for the period 1983 to 1985. Once more the incidence of energetic solar X-ray bursts and proton events fell within forecasted periods at a rate of 41 out of 46 or of 55 out of 66, depending on the strength of the events considered.[273] All results were highly statistically significant.

Predictions of solar eruptions and geomagnetic storms over 6 years yielded a 90% accuracy: 68 out of 75 events took place in the predicted time intervals. These were double-checked by independent sources.[274] It was also possible to forecast accurately the strong geomagnetic storms of 1982 and 1990. Since energetic solar eruptions have an effect on the climate via galactic cosmic rays and cloud coverage, it was possible for Landscheidt to accurately predict the end of the Sahelian drought in 1985.[275]

Harmonics of Solar System Cycles and Solar Activity

Both solar events, or geomagnetic storms occur in patterns of waves and harmonics, and they can be predicted based on our knowledge of the movement of the planets, in this instance mostly Jupiter around the Sun and the centre of mass. But we can go further: we can recognize even finer harmonic relationships if we look at the proportions of minors and majors of the golden section.

---

[273] Theodor Landscheidt, *Sun-Earth-Man*, 31–33.

[274] Theodor Landscheidt, *Solar Activity*.

[275] Theodor Landscheidt, 'Solar Oscillations, Sunspot Cycles, and Climatic Change' in Billy M. McCormac editor, *Weather and Climate Responses to Solar Variations*, 1983.

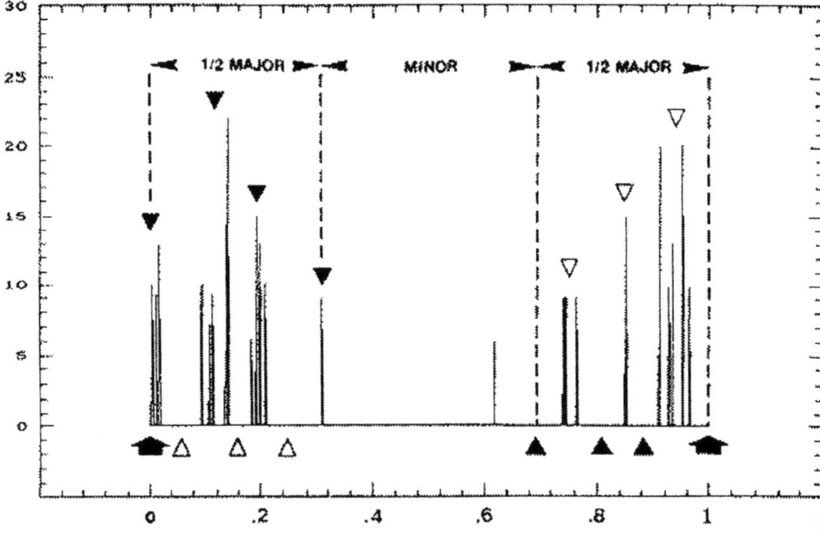

Figure 36: Energetic solar eruptions in relation to cycles of small fingers

Figure 36 shows how energetic solar eruptions occur in relation to the small finger cycle. The small finger cycle is here represented in percentages (from 0 at start to 100% at completion), and the major of the golden section is divided in two: one half after the first start of a small finger (SFS), the second half before the next SFS. We find thus a concentration of solar eruptions just after the start of a small finger and before the following one (marked by the fat arrows). To the left, immediately after the SFS the black arrows pointing downwards indicate concentration of solar activity; the empty arrows pointing upward relate to lulls of activity. If one looks more closely at a given interval of the cycle, one will see that the eruptions concentrate at the boundary of the first half of the major (here divided in two around the SFS) and practically none during the minor interval (0.382…). The patterns are reversed, they are anti-symmetric, just before the next SFS—linear increase of white triangles instead of linear decrease of black triangles— in a way that is highly statistically significant, so predictable in fact that it offers another opportunity for dependable predictions. After the publication of this figure, a further strong eruption taking place on November 6, 1997, coincided exactly with one of the active phases in Figure 36.[276]

---

[276] Theodor Landscheidt, *Solar Activity*, 17.

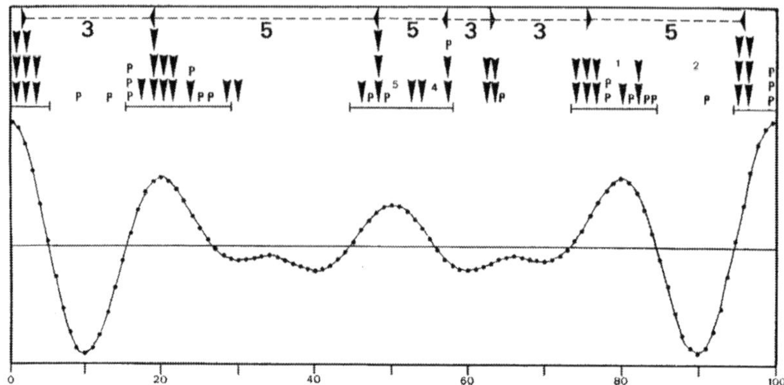

Figure 37: Composite average wave formed by the superimposition of harmonics with the frequency ratio 4:5:6.

In Figure 37 we can appreciate that no matter the length of the JU–CM–CS cycles (varying from 2 to 16 years) there is a prototypal pattern of maxima and minima at given intervals. The peaks (maxima) correspond to very energetic x-ray bursts (flares) or proton events. The clusters of these events occur mostly during the positive phase (positive or close to positive values) of the composite, average wave. The plotting was done for the years 1942 to 1969, then repeated for 1970 to 1986. The distribution is statistically very significant.[277]

The pattern of the composite wave is the result of the superimposition of the fourth, fifth and sixth harmonics, musically speaking. This indicates that the major perfect chord (4 : 5 : 6) that unites the major third (4/5) and the fifth (2/3) to the fundamental note is another basic structural element of the planetary system. This pattern was used for predictions for the cycle going from 1982 to 1990, expecting two types of intervals, one in which solar overall activity would be 2.5 times higher and another one of lull. The forecast, checked independently by astronomers and by the Space Environmental Services Center of Boulder, proved correct.[278] Here is then one more pattern that can be used as a forecasting tool.

Figure 37 shows in addition other harmonics in sections of the curve: clusters of highly energetic events form intervals of the major sixth (3:5) quite precisely, as is visible at the top of the diagram on both sides. We can also see concentration of events that

---

[277] Theodor Landscheidt, *Sun-Earth*-Man, 58.

[278] Theodor Landscheidt, *Sun-Earth-Man*, 57–58.

correspond to the octave (½) between 80% and 100% and of the major third (4:5) between 50% and 60%. Landscheidt concludes that: 'consonant intervals play an important role with respect to the Sun's eruptional activity'.[279]

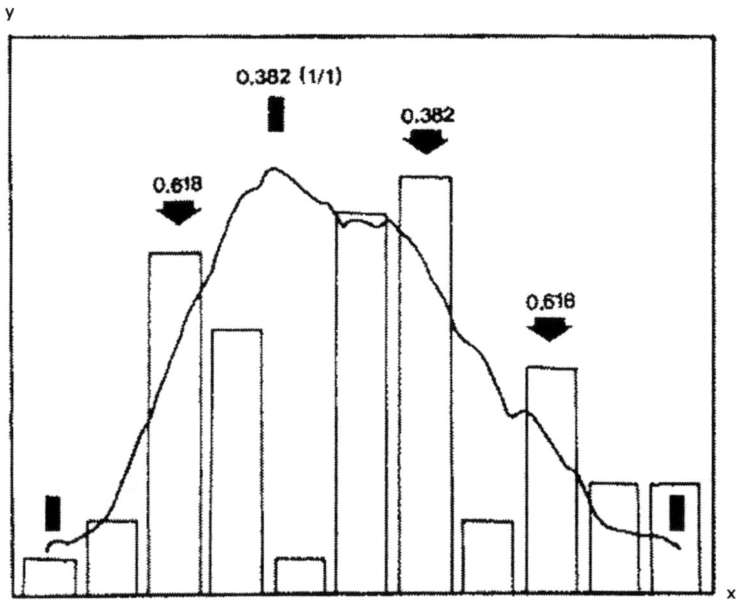

Figure 38: Distribution of energetic solar eruptions within the 11–year sunspot cycle

Similar patterns to the above extend to other parts of solar activity, which point to the importance of the golden section. Reliable data about the 11–year sunspot cycle have been gathered since 1750. What stands out from these is that the ascending phase of the cycle averages in length 4.3 years, or almost the perfect equivalent of the minor of the golden section (11 x 0.382 = 4.22). The descending side is equal to the major (see Figure 38). If we look at the distribution of strong solar eruptions within the 11–year cycle a similar pattern emerges even within ascending and descending phases. The first maximum falls on the major, the second one on the minor, and a relative maximum on the major of the descending phase.[280]

Other relationships based on the golden section also unite different cycles. The complete cycle of sunspot activity is the Hale cycle

[279] Theodor Landscheidt, *Sun-Earth-Man*, 76.
[280] Theodor Landscheidt, *Solar Activity*, 14.

of 22.1 years (twice a Schwabe cycle of 11 years) after which the Sun's activity returns to the original polarity. This value is a mean value and so is the value of the big finger cycle (BFC) of 35.76 years. If we apply the major to the latter cycle we obtain:

35.76 (BFC) x 0.618… (major) = 22.1 years (Hale)

The length of the magnetic Hale cycle and of the 11–year sunspot cycle (Schwabe) is thus connected with the fivefold symmetry in the Sun's oscillations about the invisible centre of mass of the solar system. We could say that the length of these important cycles of solar activity can be explained in astronomical terms. Landscheidt concludes: '… the stability of the planetary system hinges on the golden section, which is intimately connected with fivefold symmetry that emerges in the Sun's dynamics, which again is related to the Sun's activity'.[281]

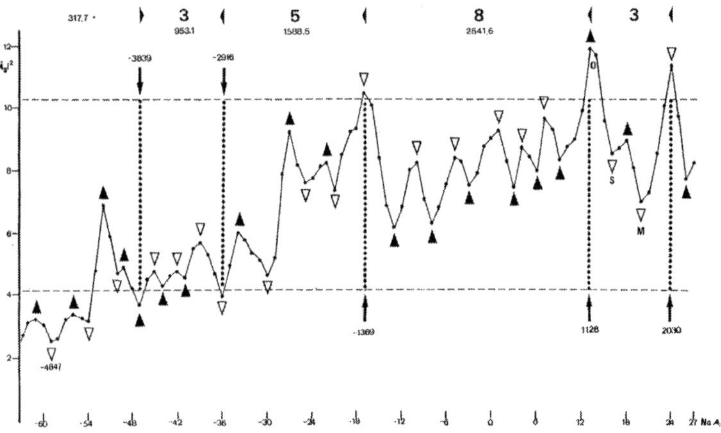

Figure 39: Combination of the consonant intervals of the major sixth (3:5) and minor sixth (5:8) in relation to the length of the sunspot cycle

We can extend our gaze over millennia only to find similar patterns once more. The intensity of impulse of torque (IOT) is measured by the change in the Sun's angular momentum. IOTs following each other in the Sun's motion around CM (the dots in the broken lines) form a sinusoidal pattern. The maxima and

---

[281] Theodor Landscheidt, *The Golden Section*, at https://plasmaresources. com/ozwx/landscheidt/pdf/TheGoldenSection_ACosmicPrinciple.pdf.

minima of these coincide with maxima and minima in the secular sunspot cycle of Figure 39. At times maxima of IOT correspond to maxima of temperature and minima to minima—as in the present phase which will extend to 2030—while at other times IOT maxima correspond to minima of the sunspot cycle, and IOT minima to maxima of the sunspot cycle, due to the phase reversals of the years -3839, -2916 (minima) -1369 BC and 1128 (maxima).

In our present phase (1128 to 2030) both secular minima and maxima of the sunspot cycle are included as can be seen by the letter O for the Medieval Optimum, S for Spoerer Minimum and M for Maunder Minimum (during the Little Ice Age). S and M both correspond to IOT minima in the graph.[282]

What is of greatest interest are the phase jumps, or reversals. These occurred in -3839, -2916 (minima), -1369 BC and 1128 AD (maxima). The jumps take place when the IOT exceeds 10 in one direction or is lower than 4 in the other. At such times we see inversion of minima and maxima relationships. From this cycle predictions can be made, such as a likely supersecular minimum around 2030. The intervals that separate consecutive phase jumps show relationships to consonant intervals, representing major sixth (3:5) and minor sixth (5:8) as can be seen with the numbers on top of Figure 34. We can also notice how threefold, fivefold and eightfold values marked in the graph are close to theoretical values. The calculated value for the interval -3839 to -2916 (first top left) yields 953.1 in relation to the actual 923; the calculated value of 1588.5 of interval -2916 to - 1369 (top middle) compares to the actual of 1547; the calculated value of 2541.6 of interval - 1369 to 1128 (top middle) with the actual of 2497. The deviations from theoretic values are of 3.2%, 2.6% and 1.8% respectively.[283]

If the above solar/terrestrial events can be predicted by relative positions of Sun and planets, so can their correlated climatic events. To these we turn now.

Climatic Effects

Landscheidt's work is truly remarkable when considered in his connections with climate. If one looks at the years 1000 to 1950 it is possible to trace very similar patterns in radiocarbon data (top), proxy data

---

[282] Theodor Landscheidt, *Sun-Earth-Man*, 79–83.

[283] Theodor Landscheidt, *Sun-Earth-Man*, 77–79.

reflecting solar activity through the square value of IOT extremes (middle) and temperature time series (bottom). The three curves coincide in highlighting the Spoerer Minimum (S ~ 1400), the Maunder minimum (M ~ 1670) and the Medieval climate optimum (O ~ 1125). The amplitude of variations of the 11-year cycle and temperatures follow very closely the changes in irradiance variations giving us a first indication of a close link between solar activity and climate.[284]

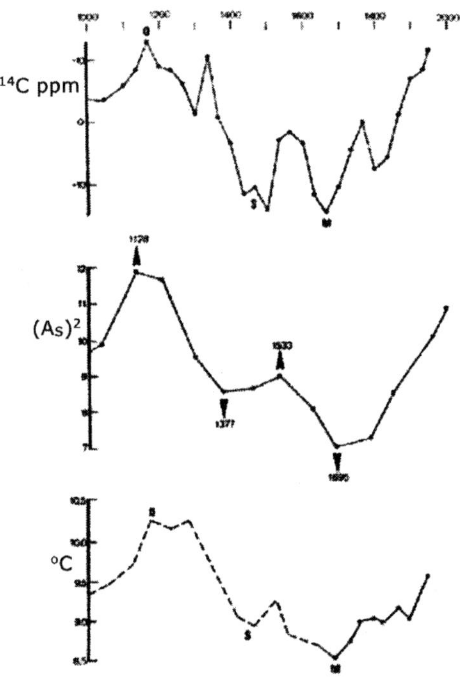

Figure 40: Temperature time series in relation to C_{14} radiocarbon proxy data (top); data derived from a semiquantitative model of cyclic solar activity (middle); proxy data reflecting solar activity (bottom) for the interval 1000 to 1950 AD

The synchronism of these three time-series over almost a millennium opens a possibility of medium-range forecasts, since the data in the second curve are based on calculations that can be extended into the future. Based on this, Landscheidt forecast in 1982 a new Little Ice Age to come around the year 2030. Many other authors at present actually concur.

---

[284] Theodor Landscheidt, *Solar Activity*, 5.

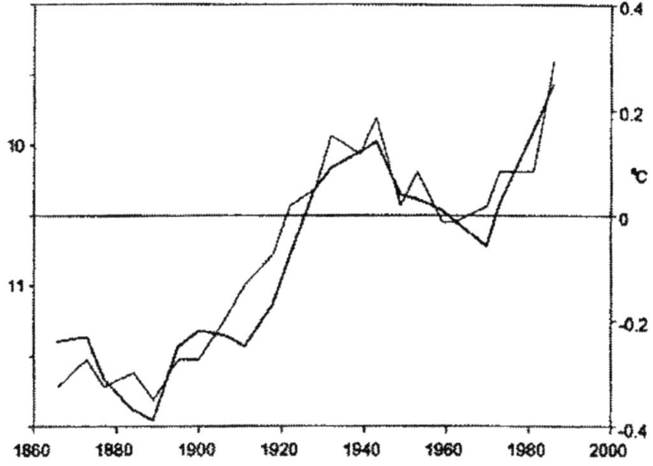

Figure 41: Northern hemisphere surface temperature in relation to the length of the sunspot cycle

Figure 41 correlates northern hemisphere surface temperature (thick line and right axis) with the length of the 11–year sunspot cycle (thin line and left axis) covering the years 1865 to 1985. Very clearly both curves' patterns match each other. Short sunspot cycles (less than the average 11 years) generate sunspot maxima, and to these correspond higher temperatures; conversely long cycles generate weaker sunspot activity and lower temperatures.[285] All of this leads to a much higher direct influence of solar activity on climate than the IPCC is willing to concede.

Another cycle that encompasses roughly 36 years was already known in 1887, the year in which it was discovered by E. Brückner. He showed this cycle's presence in diverse manifestations according to Earth regions, where it shows in synchronized phases of 33 to 37 years. We now know that these correspond to the big finger cycle (BFC) of 35.76 years, whose existence is confirmed in proxy data.

When we plot the annual mean temperature's average of the northern hemisphere in relation to the period 1951 to 1970 (above or below that average) the beginnings of big finger cycles (BFS) and big hands cycles (BHS) correspond to maxima up until the year 1933 (see Figure 42). At this point, as already encountered previously, a phase reversal sets in, where the start of big fingers correspond to minima.

---

[285] Theodor Landscheidt, *Solar Activity*, 3.

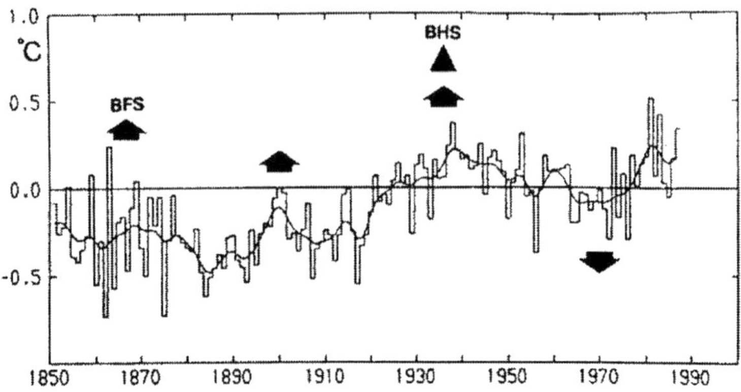

Figure 42: Annual mean temperature's average variation of the northern hemisphere in relation to the baseline period 1951 to 1970

Another confirmation of the importance of the big fingers cycle comes from correlating their occurrence to the thickness of 'varves'—deposits of silt and clay—in Lake Saki (near the west coast of the Crimean peninsula in Ukraine) The highest deposits that occur in very wet years are correlated with cycles of half big fingers (mean length of 17.9 years or ~ 35.7 / 2) at a highly significant statistical level.[286]

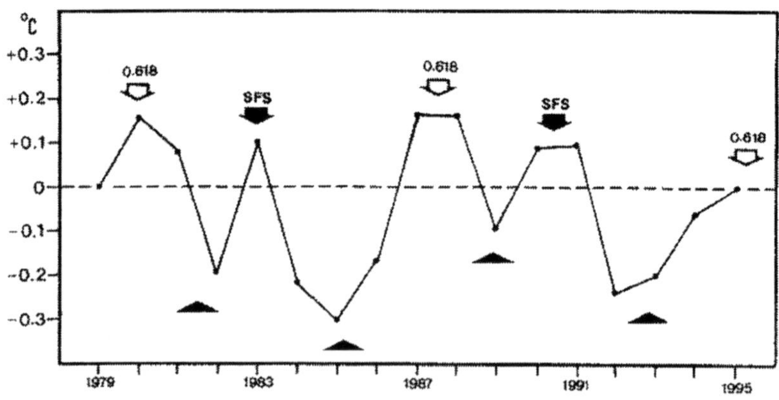

Figure 43: Global mean temperature in the lower troposphere for the 1979–1995 interval

Figure 43 relates the global mean temperature in the lower troposphere with the small fingers cycles from 1979 to 1995. The maxima

---

[286] Theodor Landscheidt, *Solar Activity*, 12.

correspond to the start of small finger cycles (SFS) and to the major (0.618). Temperature minima correspond to the midpoints between start of finger phases and majors (0.618…). Landscheidt predicted a 'middle-range minimum' in global temperature in satellite measurements for early 1997 and a maximum for mid-1998; the 1997 data showed record global low temperatures.[287]

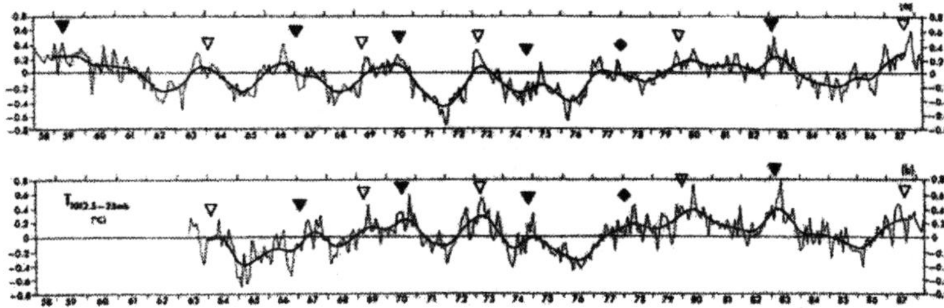

Figure 44: Global mean temperature in the lower troposphere in relation to the 1963 to 1973 mean values

Figure 44 covers the years 1958 to 1988. The curves show the monthly-mean atmospheric temperature anomalies in °C averaged over the northern (top diagram) and southern (bottom diagram) hemispheric mass between the surface and about 25–km height for the period May 1958 to April 1988 (1963 to 1988 for the southern hemisphere) in relation to the 1963 to 1973 mean values. The starts of small fingers cycles (SFS) are marked by filled triangles; the major of golden section (0.618…) within cycles formed by consecutive SFSs are marked by the open triangles. There is a close correlation between monthly mean temperature anomalies in northern and southern hemispheres and active phases of SFS (maxima) shown with the black triangles. Here again one can see peaks at the major (0.618… open triangles) and if the cycle extends over 8 years with the minor (0.382) indicated with one filled diamond. Notice that in the last two examples the impact of solar cycles is very clearly present even within the range of the last 50 years and from satellite-obtained data. So is it with the following ones as well.[288]

---

[287] Theodor Landscheidt, *Solar Activity*, 22. At the time of publication the maximum had not taken place yet and the author has not found a way to verify.

[288] Theodor Landscheidt, *Solar Activity*, 23.

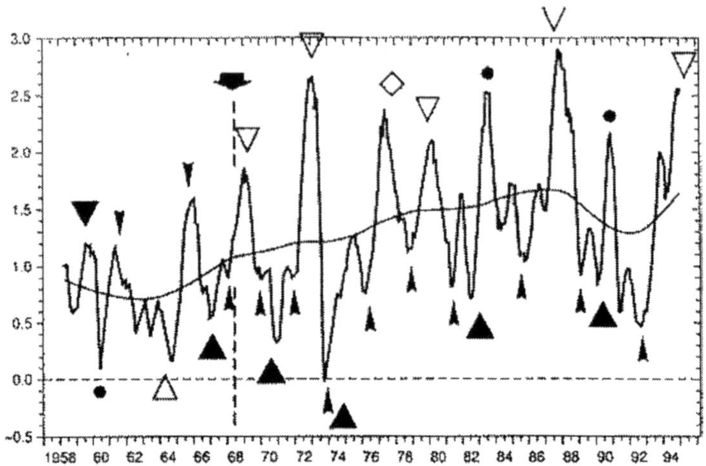

Figure 45: CO$_2$ concentrations from 1958 to 1996

Let's have a look at the pattern in the oscillations of CO$_2$ concentrations in relation to small fingers cycles in the time interval from 1958 to 1996 (Figure 45). CO$_2$ levels vary in concert with temperatures measured by satellites in the troposphere. CO$_2$ maxima coincide closely with small fingers maxima from 1958 to 1968. As in the above examples, the 1968 phase reversal, marked by the dashed line, shifted CO$_2$ maxima toward small fingers minima.[289]

Of all oceanic oscillations El Niño is the one with greatest global repercussions. The cycle of El Niño varies between 3 and 7 years on average, sometimes even longer, at which times there is a sudden warming of several °C with larger global weather disruptions. In the Figure below, if one looks at sea surface and land air anomalies during the interval 1961–1989 between 20° N and 20° S, it is noticeable that after 1968 all El Niño events peak at times of small fingers maxima. After the start of a big finger cycle in 1968 all starts of small finger cycles (SFS, open triangles) coincide with peak El Niño activity. As we have come to expect, before 1968 there was a reversed pattern.[290] Other maxima of El Niño activity show in coincidence with the major (0.618) of the cycle between two starts of small fingers. When the cycle exceeds eight years then peaks also correspond to the interval of the minor (filled diamond), an altogether rare occurrence (only 1 such value in the graph). Based on

---

[289] Theodor Landscheidt, *Solar Activity*, 21.

[290] Theodor Landscheidt, *The Golden Section*, 10.

the discovery of these cycles Landscheidt predicted El Niños for 1995 and 1998, which left academic circles sceptic because of the proximity of the last previous event in 1993. However, the predictions proved correct.[291]

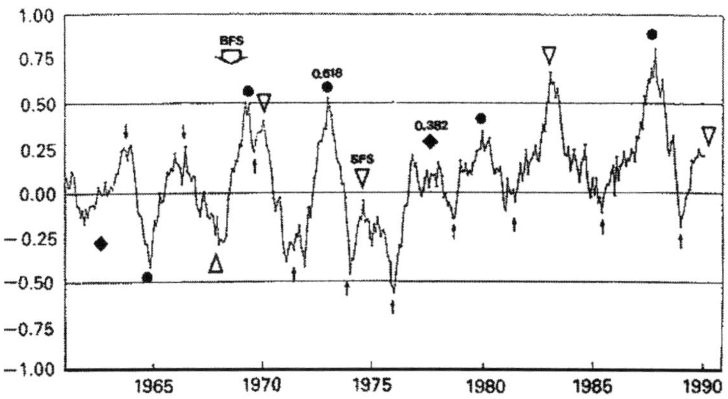

Figure 46: Monthly sea surface and land air temperature anomalies 1961–1989 for latitudes 20N to 20S

Other Connections between Solar Eruptions and Weather Phenomena

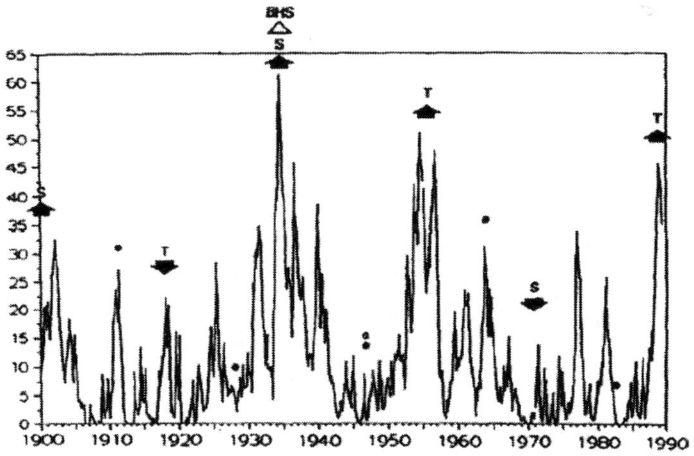

Figure 47: Palmer Drought Index in relation to BFTs and BHSs

---

[291] Theodor Landscheidt, *Solar Activity*, 21.

We can now move into the realm of precipitations or lack thereof. The Figure above plots the percentage of areas covered by drought in the US (Palmer Drought index) from 1900 to 1990. Dramatic maxima appear. Before the turning point of 1933 the start of big fingers cycles coincided with drought maxima and the tips with drought minima. The pattern was stable also after 1933 but with the usual 180° reversal.

Notice that the pattern of fingers and hands reappears at a smaller scale, so that the dip around 1920 (black arrow T) corresponds to a big finger, but smaller fingers fall even lower. Notice also the black dots in relation to the major of the golden section. Landscheidt predicted this pattern would have led to drought in 2007.[292] A confirmation of this came with the 2006–08 crippling drought that struck the US southeast.

Another pattern to watch closely is that of thunderstorm activity, which increases up to 60% after solar eruptions.[293] Here too big and small fingers correlate with rainfall amounts, from the data of fourteen German weather stations. Rainfall maxima correspond to small fingers maxima, rainfall minima with small fingers minima at a high statistical level of significance. Rainfall and temperature data from England, Wales, USA and India confirm these findings.[294]

If we connect the sunspot cycle with its maxima and minima with meteorological data here too we have clear correlations and patterns that reflect the golden ratio. This is the case if we look at the number of days with thunderstorms from 1810 to 1934 in Kremsmünster, Austria; thunderstorm frequency in Vienna from 1878 to 1934 and the number of houses struck by lightning in Bavaria between 1833 and 1879. The maxima fall on the minor and major of the solar sub-cycles.[295]

US data from 1900 to 1924 and 1925 to 1949, plotting 16,056 heavy monthly rainfalls at 1,544 US weather stations, were checked for correlations with Moon phases. While the peaks offer a clear cyclic pattern (see the superposition of solid and dashed lines in Figure 48) no obvious pattern is discernible at first sight. Yet, there is a statistically significant correlation with the

---

[292] Theodor Landscheidt, *Solar Activity*, 13.
[293] Theodor Landscheidt, *Solar Activity*, 18.
[294] Theodor Landscheidt, *Solar Activity*, 20.
[295] Theodor Landscheidt, *Solar Activity*, 15.

proportions of the golden section. Rainfall maxima correspond to the golden section major of the interval between phases (0.618) from half-Moon to half-Moon; rainfall minima take on the pattern of the minor (0.382).[296]

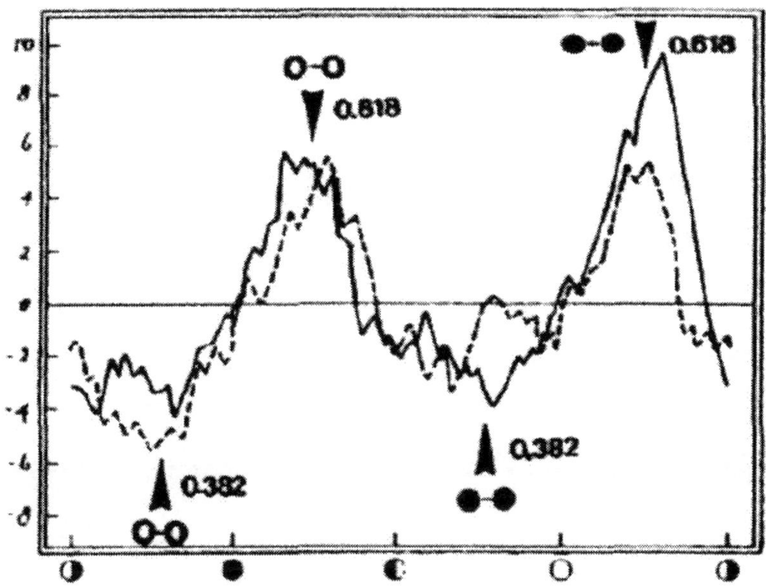

Figure 48: Heavy monthly rainfalls from 1900 to 1924 (solid line) and 1925 to 1949 (dashed line) in relation to the intervals of the golden ratio

The patterns of Moon phases lead us back to where we started (Chapter 1), only now with some additional information. The results plotted in Figure 49 show the yields of wheat seeds germinated every Friday during a 6–month period. The seedlings were measured for stem growth (length). Some data (18th and 22nd week) were lost, hence the interruptions in the curve. Here the growth of the wheat germ shows a recurrent, regular pattern during the cycle but no clear relationship with the Moon phase. The mystery is solved if we turn once more to the proportions of the golden section between two recurrent phases.[297]

---

[296] Theodor Landscheidt, *The Golden Section*, 6.

[297] Theodor Landscheidt, *The Golden Section*, 7.

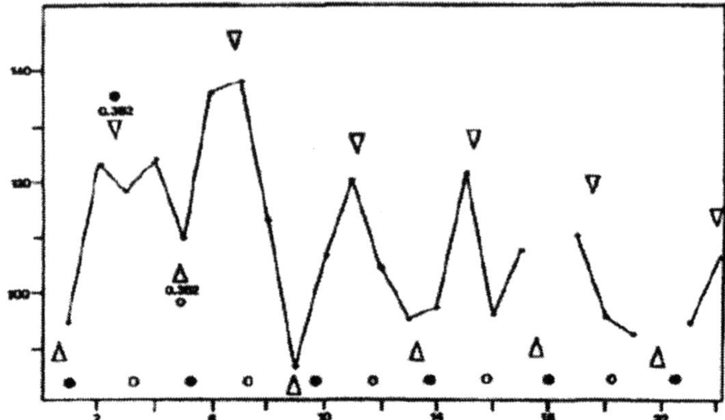

Figure 49: Wheat germination yield in relation to the phases of the Moon

At the bottom of the graph full and new Moon phases are indicated with full and empty circles. Minor sections of the interval from new Moon to new Moon (open triangles pointing downward) coincide with the peaks of yield; minor phases of the intervals from full Moon to full Moon (open triangles pointing upward) correspond to yield minima.[298] The author indicates that similar results were obtained by experiments performed by Lilly Kolisko in 1936.[299]

Astronomy and Climate Computations

Landscheidt's perception of the hidden connections between solar/planetary movements and activity on one hand, and climatic data on the other led him to accurately predict solar eruptions and geomagnetic storms, middle-range minimum in global temperature in satellite measurements for early 1997 and a maximum for mid-1998, the end of the Sahelian drought in 1985, the El Niño events of 1995 and 1998, drought in the US and various other forecasts. All of the above data correlate at a high statistical significance, the litmus test of scientific meaning. On the basis of the weight of evidence we can agree with Landscheidt that: 'climate variations are governed by the sun, not by mankind' at least at the macro-level.[300]

Instead of building abstract models, and turning their inbuilt assumptions into axioms, as we witness at present, we can let the

---

[298] Theodor Landscheidt, *The Golden Section*, 9.

[299] Original data not available to the author.

[300] Theodor Landscheidt, *Solar Activity*, 23.

cosmos guide us. Relationships exist between Sun, planets, even the farthest cosmos, and the Earth, on the basis of which we can detect and predict changes in solar activity and in earthly climate. No need to extrapolate, just recognize.

What Schauberger has shown us in relation to Earth ecosystems can be extended with an eye to the cosmos. Schauberger had so deeply intuited the functions of omnipresent polarities. It seems Landscheidt and others like him have seen these polarities at play in the larger universe. The old quadrivium of the liberal arts can now be revived on a fully scientific basis. Musical harmonics are at play in the movements of the planets. With the help of geometry and the mathematics/music of harmonics we can detect and predict the evolution of solar activity and climate to quite an accurate degree at least in the short and medium term.

Landscheidt was a modern pioneer of a new understanding of astrology and astronomy, but many bases for his work had already been laid even before him. In the last ten to fifteen years others have followed on his tracks. Such is the work of Nicola Scafetta and Antonio Bianchini, who reiterate that solar cycles are echoed in the variations of the Earth's climate and can thus be used to forecast climatic changes; and, just like Landscheidt before them, that planetary oscillations are linked to one and the other.

Continuation of Landscheidt's Work in the Present

Already in 2012 Nicola Scafetta compared the IPCC 2007 GCM model (a CMIP3 model) performance with that of an empirical climate model based on astronomical harmonics. The model used cycles of 9.1, 10–10.5, 20–21, 60–62 years, and it was tested on the interval 1850–2011. It was found to adequately forecast the period 1950 to 2011 based on the data from 1850 to 1950, and hindcast the period 1850 to 1950 based on the 1950 to 2011 data.[301]

In the model Scafetta used as a primary factor the Sun's velocity in relation to the Centre of Mass—much as Theodor Landscheidt had done—and the historical record of mid-latitude aurora events. The latter events are created and rendered visible when the solar wind collides with the Earth's atmosphere. Many premises of the

---

[301] Nicola Scafetta, *Testing an astronomically based decadal-scale empirical harmonic climate model versus the IPCC (2007) general circulation climate models*, 2012.

model's modus operandi echo and enlarge what has been said from Landscheidt's work, such as the oscillations of Jupiter and Saturn [in relation to CM] and changes they bring in electromagnetism of the upper atmosphere, or the cycles that affect cloud cover and hence albedo.

Through his model Scafetta predicted that there would be a plateauing of the temperature rise due primarily to the 60–year cycle in a negative phase. This stood in contrast with IPCC predictions of a 0.3 to 1.2 °C rise, which corrected earlier predictions of 1.0 to 3.6 °C.

Scafetta's studies and another nine published ones concur that 50 to 70% of the twentieth century warming can be attributed to variations in solar activity since the Maunder minimum of the seventeenth century.[302] Compare this with the IPCC's 2007 attribution of 90% of the warming to anthropogenic sources.

Harmonic models have evolved and improved. In 2023 Nicola Scafetta and Antonio Bianchini, both researchers of the University of Naples, have refined major planetary harmonics that correlate with the basic solar sunspot cycle (Schwabe), elucidating an important riddle, that of the deviations of the cycle from its average value. To this end they follow the work of astronomer Rudolf Wolf who, already in the nineteenth century, intuited the influences of Venus, Earth, Jupiter and Saturn on solar activity.

The two researchers speak of an Earth-Jupiter-Saturn model for the Schwabe 11–year sunspot cycle. The guiding principle of the model developed is that the smaller the difference in days between the conjunctions/oppositions of Jupiter and Saturn with Earth and the Sun, the greater is their impact on the Sun.[303] The planetary functions created by these models appear to be tightly associated with the 11–year sunspot cycle from 1700 to the present.

---

[302] Among those studies quoted by Scafetta are the following: Scafetta and West (2007), Scafetta (2009), Loehle and Scafetta (2011), Soon (2009), Soon et al. (2011), Kirkby (2007), Hoyt and Schatten (1997), Le Mouël et al. (2008), Thejll and Lassen (2000), Weihong and Bo (2010), and Eichler et al. (2009). Moreover, Humlum et al. (2011).

[303] Nicola Scafetta and Antonio Bianchini, *Overview of the Spectral Coherence between Planetary Resonances and Solar and Climate Oscillations*, 2023 at https://www.semanticscholar.org/paper/Overview-of-the-Spectral-Co-herence-between-and-and-Scafetta-Bianchini/46b65c94d3e69298625b-2447da1b8960eb0b302f

In Figure 50 we notice that the measured Schwabe cycle spectral bands fit well two tidal cycles: the 9.93–year neap–spring tidal cycle between Jupiter and Saturn (PSJ) and the 11.86–year Jupiter's orbital period (PJ). In relation to the Moon, a 'spring tide' refers to the 'springing forth' of the tide (high tide) during new and full Moon (conjunction and opposition). A neap tide—seven days after a spring tide (squares)—refers to a period of moderate tides, when Sun and Moon are at right angles to each other. The same is true for the planets, though at different cycles' length.

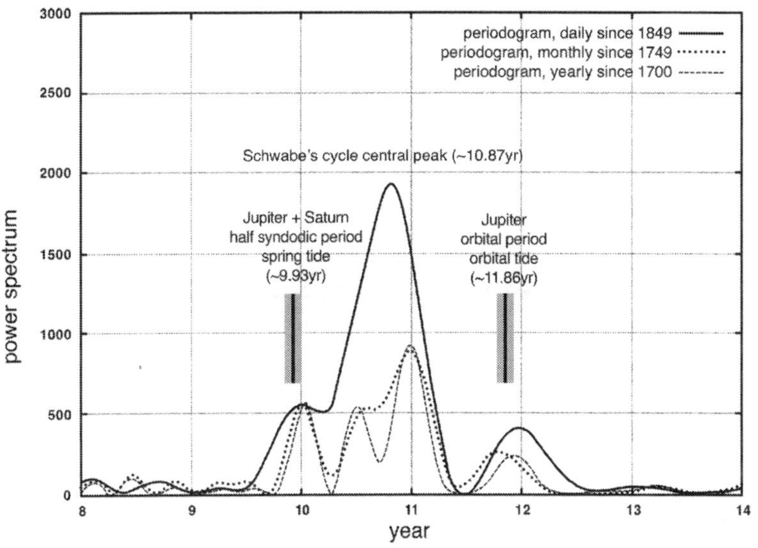

Figure 50: Sunspot power spectra (simplified from Scafetta and Bianchini)

The 11–year sunspot cycle is actually composed of three closely-spaced frequencies. It could be characterized by a primary peak close to 11 years—which could be generated by the primary synchronization of the 'solar dynamo', its internal mechanism—and the presence of the two minor peaks at about 9.93 and 11.86 years, which correspond respectively to the periodicities of the Jupiter–Saturn neap–spring tide (PSJ = 9.93 years) and Jupiter tide (PJ = 11.86 years).

The authors have discovered that out of 23 solar cycles, between February 1755 and December 2008, 11 of them had a period between 9.0 and 10.5 years—which optimally corresponds to the period of the Jupiter–Saturn neap–spring tide—and 12 of them a period between 11.25 and 13.6 years—tidal oscillation associated with the elliptic orbit of Jupiter. No sunspot cycle length was recorded with a period in the intermediate range—larger than 10.5 years or smaller than 11.25 years. The periodicities PSJ and PJ are a perfect fit for the 11–year sunspot cycle. This is an indication that the Schwabe 11–year solar cycle is primarily under the influences of Earth, Jupiter and Saturn.

Using the basic solar beat it is now possible to find harmonic equations that correlate solar activity and its oscillations to the larger climatic patterns. It is known at present that solar oscillations go from monthly to multi-millennial. The most documented ones beyond Schwabe (11 years) and Hale (22 years) cover 40–45, 55–65, 80–105 (Gleisberg) years. Other cycles cover 150, 170–240, (Jose and Suess de Vries cycles), 800–1200 (Eddy cycle) and 2000–2500 years (Bray-Hallstatt cycle).

Based on the above data, Scafetta developed a 'multiscale harmonic solar and climatic model' based on the Jupiter–Saturn neap-spring tidal oscillation (1), Jupiter's orbital tidal oscillation (2) and the solar dynamo cycle (3). Four primary beats with periods of PS13 = 60.95 years, PS12 = 114.78 years, PS23 = 129.95 years, and PS123 = 983 years result from the combination of the three cycles. The resulting model replicates the major secular solar activity minima and maxima, such as the Oort, Wolf, Spörer, Maunder, and Dalton grand solar minima, which are recorded in all solar proxy records including those of the cosmogenic isotopes ^{14}C and ^{10}Be. Based on this accurate hindcasting model Scafetta forecasted a grand solar minimum between 2015 and 2045. Others predicted the same from similar models in 2013 and 2015.[304] In this they independently agree with Landscheidt.

In addition to the above a global surface temperature model developed by Scafetta, based on astronomical oscillations, was successful in predicting two El Niño events of 2015–16 and 2020. The researcher concludes: 'the idea that changes in solar activity might be only governed by internal dynamo mechanisms appears to still fall short of fully explaining why all of these solar cycles

---

[304] See Salvador, R.J. *A mathematical model of the sunspot cycle for the past 1000 years*; Mörner, N.A. *The Approaching New Grand Solar Minimum and Little Ice Age Climate Conditions* and Courtillot, V.; Lopes, F.; Mouël, J.L.L. *On the Prediction of Solar Cycles*, quoted in Nicola Scafetta and Antonio Bianchini, *Overview of the Spectral Coherence between Planetary Resonances and Solar and Climate Oscillations*.

coincide with the most significant planetary periods derived from the planets' orbits, spring tides, synodic cycles, and invariant inequalities.'[305]

The researcher offers the following diagram of correlations that recapitulates much of what we have seen in this chapter and in Chapter 7.

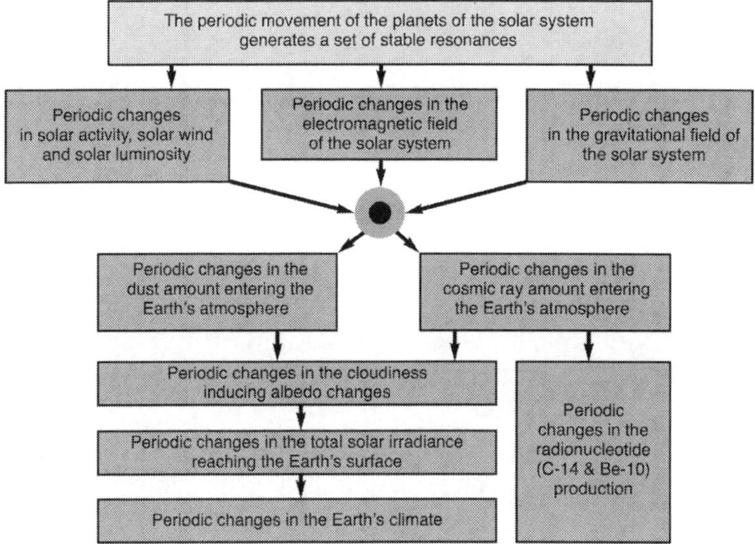

Figure 51: Planetary beat processes and the spectrum of terrestrial variables affected

Other, more complex models have evolved similar to this one at present.

The Approaching New Grand Solar Minimum

Looking at the compendium of present-day research, Nils-Axel Mörner shows that the coming of the new grand solar minimum can be inferred by many researchers from a variety of starting points:

- the phasing of sunspot cycles;
- the oceanic oscillations of the North Atlantic over the past millennium;

---

[305] Nicola Scafetta and Antonio Bianchini, *Overview of the Spectral Coherence between Planetary Resonances and Solar and Climate Oscillations.*

- the 'patterns of cosmogenic radionuclides' (^{14}C and ^{10}Be) in natural terrestrial archives;
- the Sun's movement around CM;
- the planetary conjunction history (in particular those of Saturn and Jupiter);
- the general planetary-solar-terrestrial interaction;
- the planetary spin-orbit coupling: the exchange of momentum between angular and rotational velocities of the Earth or other planetary bodies.[306]

A word about the last of the items listed above. Mörner finds that the record of past solar minima and maxima correlates respectively with periods of speeding up and slowing down of the Earth's rate of rotation. Maxima took place in mid sixteenth, early and late eighteenth, early and late twentieth centuries. Minima in mid fifteenth, early and late nineteenth centuries. 'During the last three grand solar minima—the Spörer, Maunder and Dalton Minima—global climate experienced Little Ice Age conditions. Arctic water penetrated to the south all the way down to Mid Portugal, and Europe experienced severe climatic conditions ...'

Various authors agree that the new grand solar minimum will occur somewhere between 2030 and 2040. All of this concordance of data runs once more against the IPCC narrative of a lack of evidence of solar/planetary influences on climate.

With all of the above explorations we have completed the attempt to better understand the forces that truly shape climate. We have found them in the larger ecology of our solar system. In doing so we have entirely moved away from the assumptions built into and conclusions derived from IPCC models. We are just left with the attempt to point out the foundations of a future astronomy, prerequisite for an understanding of climate change/evolution.

Limitations of Galileo's and Newton's Astronomy

Kepler's astronomy suggested the study of the solar system through harmonic and musical laws, and as we saw, Landscheidt followed this route, vastly acknowledging his predecessor.

---

[306] Nils-Axel Mörner, *The Approaching New Grand Solar Minimum and Little Ice Age Climate Conditions* (2015), at https://file.scirp.org/pdf/NS_2015111916552083.pdf.

Newton arbitrarily focused on centripetal/gravity forces, at the expense of centrifugal/levity ones.[307] The Newtonian theory of universal gravitation states that:

$$F = \frac{GMm}{r^2}$$

where F is the gravitational force exerted by the Earth upon the falling body; m is the mass of the body, M the Earth's mass, r measures the distance from the centre of the Earth to the body in question and G is a constant. Newton could only formulate this law by arbitrarily choosing some aspects of physical phenomena and ignoring others.

A more encompassing view of the matter is expressed by Steiner thus: 'However we look to the phenomena of the heavens, we must recognise that we cannot study them simply according to the laws of centric forces, but that we must regard them in the light of laws which are related to the laws of centric forces as is the sphere to the radius… It will then become apparent … that we need: in the first place, [what] has essentially to do with centric forces, and secondly, in addition to this system, another, which has to do with rotating movements, with shearing movements and with deforming movements.'[308] Newton only considered centric forces—those directed to a centre—ignoring those directed toward the periphery. However, no circular or elliptic motion, from a slingshot to a planetary movement, can be described with centric forces alone.

The whole is more fully explored from within a Goethean perspective by Gopi K. Vijaya. Starting from the basics of a uniform circular motion, the velocity v of an object moving around a circle

---

[307] Gopi Krishna Vijaya, *Replacing the Foundations of Astronomy* at https://ia802904.us.archive.org/24/items/replacing-the-foundations-of-astronomy-vijaya-gopi-krishna-2/Replacing%twentiethe%20Foundations%20of%20Astronomy%20%28Vijaya%2C%20Gopi%20Krishna%29%20%282%29.pdf

[308] Rudolf Steiner, Third Scientific Lecture Course, Astronomy Lecture 10, Stuttgart, 1921, quoted in Gopi Krishna Vijaya, *Celestial dynamics and rotational forces in circular and elliptical motions*, available at https://reciprocalsystem.org/sites/default/files/2023–09/Celestial%20Dynamics%20and%20Rotational%20Forces%20Jan%202019%20%28Vijaya%2C%20Gopi%20Krishna%29.pdf.

is perpendicular to the radius R of the circle and constantly changes direction. In Figure 52 the angular velocity ω is constant and equal to v/R.

Mathematically speaking circular motion can be derived from an infinite system of simultaneous derivatives, acting both toward and away from the centre, along and opposite to the velocity. In the more inclusive view of circular motion that Steiner suggests, to the central acceleration (pointing to the centre) are added rotational forces. All in all we have four sets of forces that support circular motion:

- toward the centre: centripetal forces;
- opposite to velocity: retarding forces;
- away from centre: centrifugal forces;
- in the direction of velocity: quickening forces.[309]

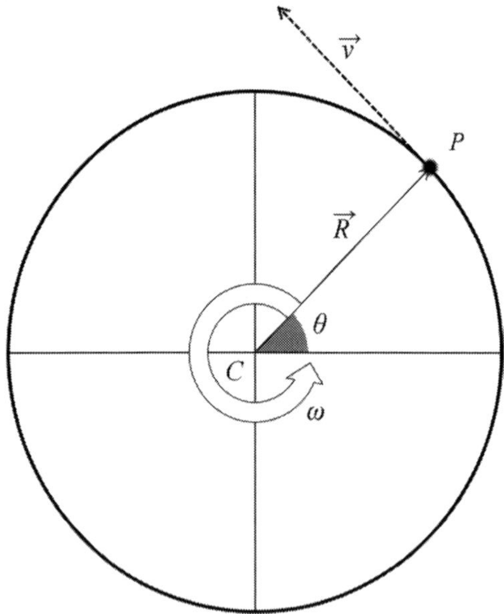

Figure 52: Uniform circular motion of an object P around the centre C

This more encompassing view of circular motion clearly shows us that it is actually impossible to produce uniform circular motion by making recourse to a single force, as Newton did. Furthermore,

---

[309] Gopi Krishna Vijaya, *Celestial dynamics and rotational forces in circular and elliptical motions.*

rectilinear and rotational motions cannot possibly be compared, nor treated in a like manner.

In elliptic as in circular motions it is actually impossible to calculate all the forces that generate and maintain the motion because π (Pi)—the ratio of a circle's circumference to its diameter—is a transcendental number, leading to an infinite number of radial and tangential forces. G. K. Vijaya concludes: 'Rather than attempting to reduce circular motion to an infinite series of linear accelerations and higher-order forces, it is preferable for astronomy to deal with circular motion on its own terms, in a descriptive fashion. Circular motion does seem to be irreducible, and if one is not able to conceptually apply current astronomical theory accurately to the moon—the simplest circular motion that we perceive—then it is impossible to do so for the other planets.'[310]

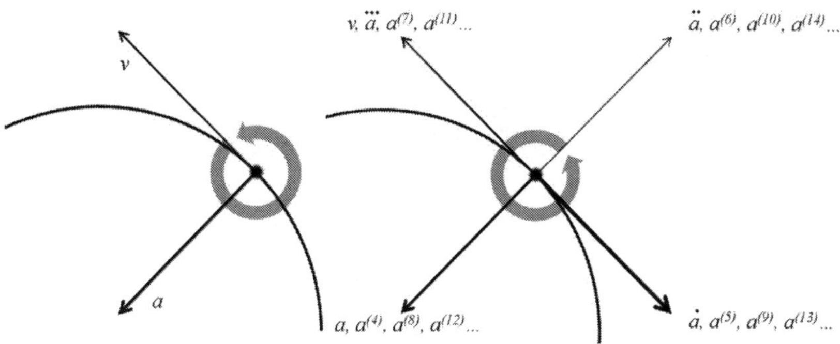

Figure 53: Radial and tangential successive rates of change

We come therefore to the sobering realization that, even if climate harmonic models can have a decadal or multidecadal predictive value, in the final analysis, we cannot replace mechanistic models with harmonic models without taking into account that it isn't possible to equate the solar system to a celestial clockwork. It is a living system, as Landscheidt aptly demonstrated, and therefore continuously evolving in ways that cannot be fully charted.

We have come full circle to a deeper understanding of the causes of climate from the perspective of the Earth (Part I) and

---

[310] Gopi Krishna Vijaya, *Celestial dynamics and rotational forces in circular and elliptical motions.*

from that of the solar system (Part II). We are left with two tasks to bring this exploration to a conclusion. We will outline the cultural revolution from which we can start to establish a firmer foundation for the understanding of climate. And we still have to approach the centre around which we have been circling, so to speak. Let us turn to the heart of the matter in our conclusions.

Conclusions and New Beginnings

'At rest, however, in the middle of everything is the Sun.'

Nicolaus Copernicus

'The Sun, the hearth of affection and life, pours burning love on the delighted earth.'

Arthur Rimbaud

In Parts I and II of this work we have looked at sets of polarities. At the level of global ecology we saw the importance of gravity and levity, negative and positive temperature gradients, technological or planetary motions. At the level of the solar system we saw the polarities of Sun and Earth, Sun and larger planets, centre of mass and centre of the Sun, major and minor (golden section). These are of course just the main ones.

We have seen, as it were, climate from two directions. At the ecological level it is not the cycle of CO_2 that plays a paramount role but rather that of water, and the results of it are cause for great concern. They have long been ignored, and they find little place in modern-scientific thinking. At the solar/planetary level it appears there is no longer reason to believe in a CO_2 threat, when everything points to much larger cycles than those of the last 150 years, and to the role of water dwarfing that of CO_2 via Sun and planets. The common element to the two approaches arises in the importance of water for the living Earth. Could this be a real surprise?

From the earthly perspective the pervasive change of the hydrological cycle has seen and unseen effects. The seen are worrisome enough and still underestimated, or mostly ignored. The unseen, such as the etheric depletion of rivers, forests and farmlands are very consequential on their own. Echoing Schauberger we can remind ourselves: 'Opposition alone, however, achieves nothing. Our youth will achieve any practical success in their struggle only when the causes are identified and the errors are revealed...'[311] We truly need to think an octave higher to address what we have called planetary depletion.

---

[311] Viktor Schauberger, *Our Senseless Toil*, 1933, quoted in Alick Bartholomew, *Hidden Nature*, 259–60.

Under this lens the real change in terms of energy lies in how we think of temperature and motion, and which way we decide to go: more of a technological treatment of all the components of the ecosystem, or a complete change toward their holistic integration with the critical understanding of the being of water at the centre. With an understanding of this critical element there is little that present climate change policies will address. In fact they will aggravate the ecological crisis by keeping us on the track of positive temperature gradients and mostly technological motion.

On the other hand we can draw a sigh of relief. We need no longer point to the 'natural variability' of physicists and modellers, when climatologists, ecologists, oceanographers, astrophysicists and other researchers have accumulated evidence of a planet in a large web of interlinked cycles embracing the whole solar system, which cause clear and understandable oscillations, not 'random variability'. Here we can actually be moved to awe and realize that we need to review simplistic, deterministic ideas of causation that fall completely short of reality.

Whether we look at climate change from the Earth or from the solar system, there is an overlap at one critical place: the Sun, the heart of our life on Earth. Just as modern science thoroughly misunderstands the heart—treating it as a pump—so does it completely miss the basic ideas for coming close to a living apprehension of the heart of our solar system. In relation to the first assertion Schauberger unequivocally sides with much of modern-anthroposophical research about the heart.[312] 'The movement of blood, however, is

---

[312] The long-entrenched view of the heart as a pump has been definitively challenged in modern scientific terms. Branko Furst, MD, in his monograph and several peer reviewed articles, adds conclusive evidence that, rather than being a pressure-propulsion pump impelling the blood through the billions of capillaries—an impossible feat for a small organ the size of a human fist—the heart merely 'lifts' the blood from a low-pressure, venous system to a high-pressure, arterial system. This is confirmed by the fact that the heart expends over 80% of its high energy consumption to generate warmth, and only a fraction (10–15%) on beating, that is, on ejection of blood into the aorta and pulmonary artery. Conceptually, rather than a pump, the heart functions more like a hydraulic ram which is powered by flowing water and converts the kinetic energy of flow into pressure. See Branko Furst, *The Heart and Circulation: An Integrative Model*, second edition.

not the result of the so-called beating of the heart—it is the cause.'[313] Not surprisingly he sees the Sun as a heart of the universe.

A New Understanding of Sun and Solar System

At the intersection of Part I and Part II of this book we find the Sun, to which we can look with new wonder. Therefore we will bring matters to a close by first attempting a tentative, new look at the being of the Sun before we return to the view from the Earth. How this reality will be perceived may determine much of humanity's future.

Step by step a wholly different view of the Sun and its relationship to Earth and climate is starting to emerge. We can start with the most widely known pieces of information. The Sun's diameter is 110 times greater than the Earth's diameter. Conventional science holds that the Sun's temperature is of 6000°C at the surface and 20,000,000°C at the centre, but this is of course an indirect measurement that poses challenges. Georg Blattmann and Callum Coats contend that we may be looking at kinetic energy—what in fact is measured—and that the translation into heat is an assumption. Kinetic activity is the type of motion that produces either heat or cold. Thus it may actually be icily cold in accordance with some of Schauberger's challenging views.

Among the many shortcomings of modern science Schauberger counts: 'the generally held view that the Sun is a molten, incandescent fireball with extremely high temperatures'.[314] In *Our Senseless Toil* he countered this negative assertion with a positive view: 'The closer we approach this source of light and heat, the colder and darker its face will become. The nearer we are to it, the brighter the stars will be and as its light diminishes, heat, atmosphere, water and life will also disappear.'[315]

---

[313] See Viktor Schauberger, and Callum Coats, editor, *Nature as Teacher*, 140, and Viktor Schauberger, and Callum Coats, editor, *The Fertile Earth*, 84. For a spiritual scientific confirmation of this point see Branko Furst, *The Heart and Circulation: An Integrative Model.*

[314] Viktor Schauberger, Callum Coats, editor, *The Water Wizard*, 77.

[315] To confirm the views about the darkness of the Sun, we know that visibility of the stars is a function of the thickness of the atmosphere. Once they left the Earth the astronauts could not see any stars, not at least until they had reached the Moon's atmosphere. Thus the invisibility could also

How can we get closer to this question from a phenomenological-ical perspective? One key phenomenon to unlock the riddles of the Sun's being are the eclipses. Over a hundred years, or slightly more, scientists have been able to accumulate systematic observations of little over two hours of the Sun's corona. The white metallic-looking sheen of the corona stretches its arms far into space. The shape of the rays of the corona—although remaining constant over the time of an eclipse—changes over time, particularly in relation to the phases of the 11–year sunspot cycle. It is rather evenly distributed around the Sun at the time of sunspot maximum. At sunspot minimum it narrows down with maximum strength around the equator and much lower intensity at the poles. Fortunately, a devise called the coronagraph has allowed to record indirect observations outside of eclipse times.

The Sun can be equated to a field of action of which the orb is only a reference point. In the reverse of point and periphery perspective that we have followed in previous chapters, the real, etheric, nature of the Sun finds itself wherever its effects are felt, echoing the intuition of the philosopher and poet Friedrich W. J. Schelling: 'The sun does not shine only where it is, it is also there where it shines.' To this view that the Sun is everywhere where its actions are felt modern-scientific observations lend weight. The veil of the corona penetrates through the whole planetary system and farther beyond Saturn according to some scientists. And satellite and space capsules seem to confirm this assertion. Georg Blattmann concludes: '… this central star is coextensive with the bodies that belong to it. The Sun is as big as the planetary system.'[316] The views that have emerged in this chapter about the solar wind as a pervasive influence can be extended to posit that this is not just a remote influence but a presence, naturally more etheric than physical.

All of the above is confirmed by another exceptional phenomenon, that of the 'zodiacal light' which is best observed at the tropics. Before sunrise or shortly after sunset one can see at times a narrow pyramid of light, with a blunt apex uppermost, which rises above the horizon and is about as bright as the Milky Way but more

---

apply to the Sun. The Sun could indeed be dark. From the Earth visibility ends where the hydrogen concentration—that is all that is left in the upper atmosphere—equals the concentration in interstellar and intergalactic space (from Callum Coats, *Living Energies*, 78–79).

[316] Georg Blattmann, *The Sun*, 34.

mist-like. On certain special occasions it is also possible to notice a weaker counter-image in the opposite direction. And even more rarely it is possible to see a luminous connection between both pyramids of light across the sky. The axis of these pyramids follows the zodiac. The halo of light fills the space within the plane of the zodiac as far as the most distant planets. The material is thickest around the Sun itself. The phenomenon indicates that the space between Sun and Earth isn't the simple vacuum it is believed to be, but is filled with a medium that reflects the sunlight. What allows the light to be refracted could be the very diluted dust derived from disintegration of planets, colliding planetoids or meteorites. Blattmann concludes: 'It is evident to us that the corona constitutes the essential nature of the sun, and as we have defined it, fills all space around the sun's orb, including the most distant planets.'[317]

As we saw, what physicists called temperature are indirect measurements of kinetic energy. This amount of energy decreases toward the centre of the Sun, (increases according to conventional science) indicating that the Sun exists most strongly in the periphery than in the centre, a phenomenon that moves in the direction of Schauberger's assertions.

Let us follow Schauberger, very much in line with Blattmann, further into what appears as unprovable, but interesting intuitions at this point: 'The superconduction of electricity, that is, the resistanceless transport or propagation of energy, takes place at extremely low thermal temperatures.'[318] We may have to do with a phenomenon of cold fusion, in keeping with everything Schauberger says about the generation of life through positive temperature gradients, in this case much lower than what takes place on Earth. The release of energy would take place through the cold fusion of hydrogen into helium.

Moving further along these lines, which reverse conventional scientific thinking, the Sun could act as the field of energy which draws the heat out of the Earth, and the heat radiation thus generated would be reflected back by the atmosphere. As the air becomes more rarefied in moving up the atmosphere there would be less reflection; heat would decrease and cold increase until we reach the limit of the atmosphere, where the concentration of hydrogen, the sole gas left, equals that of hydrogen in the interstellar space.

---

[317] Georg Blattmann, *The Sun*, 41.
[318] Quoted in Callum Coats, *Living Energies*, 81.

To contradict the temperature theory—and the physical action of the Sun rather than its etheric nature—we have also the phenomenon of heat diminishing with height in the mountains, though in theory we are getting closer to the Sun, and also the presence of extreme cold regions within the Earth's atmosphere—e.g., -60⁰C just above the tropopause, or -90⁰C at the mesopause. Since the interstellar space, which is at a near absolute vacuum, registers quite logically a temperature of -273⁰C (absolute 0), how is the Sun's heat supposed to reach the Earth?

More food for thought and open questions have emerged in the work of Landscheidt in the relationship between the Sun's etheric nature and the little of gravity that is nevertheless present in the solar system. We have seen it in the dynamic between CS (centre of the Sun) and centre of mass CM (barycentre of the solar system). Here very likely—in the balance between the two ends of the continuum—gravity, though still present, falls as its lowest, possibly the 4% predicated by Schauberger, levity to its highest. The work of the German pioneer shows us that the Sun's orbital motion around CM varies according to the small variations in centripetal and gravitational forces.

We can posit that in the perpetual creation of boundary conditions that take place around the Sun, there are two such boundary surfaces—presently also called 'attractors'—on one hand the Sun's corona, on the other the centre of mass, CM. Upon CM works all of the solar system's gravity, especially that of Jupiter and larger planets; CS, the Sun's centre within, is the unattainable centre of levity.

Boundary phases are linked to instability, which lead to the spontaneous formation of new structures. At the boundary transition from one polar quality to the opposite one, instability will arise together with new patterns. The Sun's surface is just such a cosmic boundary, separating two attractors: 'one of contraction, where dying suns transform into neutron stars or black holes; one representing expansion and dissipating radiation'.[319] This gives rise to the solar activity of sunspots, faculae, eruptions and flares. Energy emanating from the boundary layer of the Sun meets with another boundary layer at the surface of the Earth: in essence everything we have seen through the work of Schauberger and other Goethean or spiritual-scientific authors.

---

[319] Theodor Landscheidt, *Sun-Earth-Man*, 13.

To follow this exploration further we will turn to the work of Gopi Krishna Vijaya, with the advantage of a precise mathematical and physical treatment of the matter, that reaches further than the intuitions of Schauberger and Landscheidt. To move into this new direction we will first start with some unique observations that have been known to science but have not raised the necessary wonder, nor moved science to the conclusions to which they point.

It is to say the least astounding that it takes both Moon and Sun close to 27.32 days to achieve a complete revolution around their axis. The difference is minimal. The parallels don't stop here. The phenomena of the eclipses show us that Sun and Moon appear to the naked eye as celestial objects of the almost exact, same size. They form a visual angle of ½ degree each or 1 degree together. The number 360, already known to old cultures is thus not an arbitrary choice, but an ultimate reality.

To human beings of the present, time is perceived in the same manner as space, for example through the movement of the hands of a clock; in other words it is converted into space. However, this perspective ultimately only holds true within the realm of Earth and of gravity. In terms of physics, once we move within the solar system and cross the threshold from gravity to levity the terms change. G. K. Vijaya explains: 'Once this limit is crossed, the nature of space changes from linear to angular; therefore one is no longer looking at a linear extension or *distance*, but instead its angular equivalent: *a solid angle*. This angular measure alters the vectorial nature of space, where there is no longer a specific direction that one can point to.'[320] In his work we find expressed in physical and mathematical terms the reality of 'boundary surfaces', or 'attractors'. The Sun is one such boundary surface. In other words the order of reality that we know within Earth boundaries crosses over into another one at the boundary of the Sun. We can no longer extrapolate what we know of one order of reality into the other. We cannot apply the logic of Earth to the farthest reaches of space, as modern astronomy does. We need to completely adapt and reverse our habitual thinking.

---

[320] Gopi Krishna Vijaya, *Unlocking Eclipses and Planetary Distances*, available at https://www.gopivijaya.com/_files/ugd/01c972_441e597e-09984c37a2647a1a43939681.pdf

Continuing the thought of the reversal of centre to periphery that we have explored in Part I of this book we can contrast Sun and Moon according to gravity and levity in relation to light and how it travels. Whereas the light of the Moon is received radially, streaming to the periphery from a centre, that of the Sun is received peripherally (see Figure 53). We have here a relationship between a line and a surface: 'In other words, the radial distance to the Sun measured is not the real distance at all, but the hemispherical surface area *taking on the appearance of a distance.*' And further: 'The relation between the Sun, Moon and Earth hence unlocks the distance-relations in the Solar System.'[321] This indicates that this relationship holds true as well for the other planets of the solar system.

Seen pictorially, we can contrast the radial with the peripheral travel of light in the contrast between a line and the surface of a hemisphere on the other, with the Earth at its centre, Moon and Sun on either side (see Figure 54). While to our sensory perceptions Sun and Moon show many similarities, the same apparent size and rotational periods, this is because they are like two sides of the coin, corresponding to two polar sets of realities.

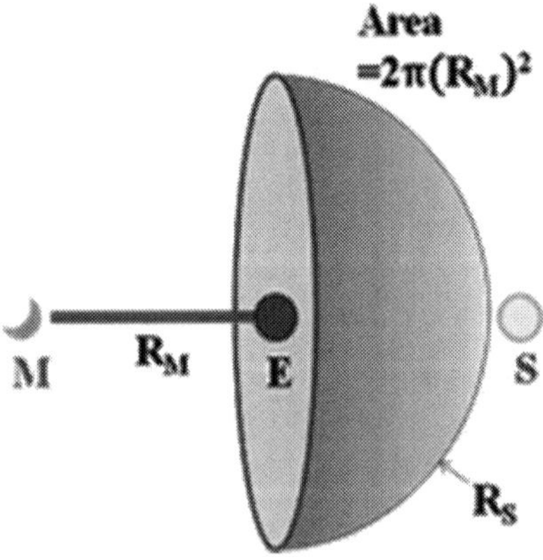

Figure 54: Moon-Earth-Sun System

For the sake of our exploration it is not necessary to treat the matter mathematically to its ultimate consequences. For this we refer the interested reader to the work of G. K. Vijaya. Suffice to

[321] Gopi Krishna Vijaya, *Unlocking Eclipses and Planetary Distances.*

mention that when all is said and done Kepler's astronomy, particularly his Harmonic Law, is amply vindicated over and against the modern view of astronomy that we have inherited from Galileo and Newton.[322]

The above leads to the conclusion, in Vijaya's words, that: 'the Sun forms a unit boundary for the Earth, and not merely a "zero point of the reference system". The exact same shift that has to be made from the conventional reference system to one based on unit speed, has to be made in astronomy from the conventional "Sun-centered" calculations to treating the Sun as a *unit boundary*: the gravitational limit. And just as a sphere has an inner surface and an outer surface, this particular unit boundary has the Moon on one side, and the Sun on the other side of this boundary.'[323]

This work has come to the unescapable realization that the Earth's climate is influenced by the whole planetary system. What modern science has increasingly demonstrated since the nineties has now expanded to bring us back to the views that Pythagoras held almost two thousand five hundred years ago. We may stand at the cusp of the modern discovery of the harmony of the spheres, one that is presently based on accurate observation and calculation, and rigorously formulated empirical equations. That it has some solid foundations is proved by the fact that it can predict aspect after aspect of climate on the basis of heavenly phenomena alone, at least in the short and medium term. Future models need only use the equations of the heavens, the movements that can be perceived and recorded directly, and predicted and verified. However, we need to take into account that this can only be done for the relative short or medium term, after which the universe claims its due mystery and the human mind must rest content with its limitations.

---

[322] Kepler's Harmonic Law states that changes in time (t) vary according to the second power (t^2), and changes in planetary distance (r) vary according to the third power, or as a volume (r^3). The formula, applicable to all planets is $r^3/T^2 = (1\ AU)^3/(1\ year)^2$, with AU, the astronomical unit equal to the distance between Sun and Earth. Through this formula we obtain values of r^3/T^2 all very close to 1 for all nine planets, Earth included.

[323] Gopi Krishna Vijaya, *Unlocking Eclipses and Planetary Distances*.

A New Understanding of Gaia

The tragedy of Schauberger's life is that he came at a time in which he could only be a prophet for the future. What he had to say simply could not take root in the time in which he lived, no matter how much hope his views and practical alternatives can generate. That was the hardest trial of his life: failing to understand that he actually did accomplish all that was possible in his time and accepting that he had broadcast many seeds that will need a new culture in order to blossom. The old culture will have to die the hard way before these seeds will germinate and grow. Until then it will do everything in its power to hide, condemn and malign all viable alternatives.

In Schauberger's views, as in many of those who embrace a Goetheanistic and/or a spiritual-scientific understanding of the world, matter is a by-product of energy, or of spiritual reality. In Schauberger's words, everything we have seen so far, could be summarized thus: 'All natural systems are mirrors of their pattern of energy, or of the "idea" that sought to create them in the first place. When the system is in place, the energy from which it originates is rejected as matter being too coarse to be carried further in the energy stream.'[324]

This view can be pushed further to an understanding that the quality of our life, of our health and thinking depends primarily on what kind of spiritual nourishment and etheric energy circulates through us, more so than on what materials—food, air, water, natural environment—we ingest or come in contact with. This is not a dualistic view of the world. When seen closer we could say that it is not sufficient to have access to food, water, air, natural environment … if we do not pay attention to what spiritual forces are still present in all of these elements.

Nature, the Human Being and the Spirit

It is well understood among biodynamic farmers that it is not sufficient to advocate a return to the past and produce *foods that do not have* a whole host of offending materials—herbicides, insecticides, hormones, trace pollutants, etc. It is *what is in the food* that renders it fit for human consumption to such an extent that it supports the spiritual in the human being. Hence the importance of an agriculture which takes cosmic rhythms into account and ennobles matter

---

[324] Alick Bartholomew, *Hidden Nature*, 82.

through careful rotation of crops, biodynamic preparations, operations timed with the rhythms of the Moon and its place in the signs of the zodiac, and so forth.

This view was once again a matter of direct perception for someone like Schauberger; witness this as one of many statements to the effect: '"A man is what he eats." The higher the quality of the food he eats, the nobler the products (thoughts) of the energies created in the digestive system.' This led him to conclude:

> Today people have become so destitute and have been so intellectually castrated through perverted methods of food production, that they will now have to dance around the Golden Calf. They can see no other possibility of regaining their lost freedom and personal independence other than through the enforced redistribution [communism] of the necessities of life that are still available. Should they do this, however, then they will find themself in slavery of a hitherto unknown order.[325]

This was very similar to what Rudolf Steiner thought as well.[326]

From the above we can understand that the quality of what we bring to the world through thoughts, feelings and actions depends on the quality of energy that we ingest and surround ourselves with, starting from the energy in our water and in our foods. This is why Schauberger was an advocate of high-quality water and nutrition. He saw the importance of the physical/energetic support to the brain. Otherwise, our thinking, little supported by the brain, would only be able to think logically, but not biologically, not holistically. This is a conundrum which is part of a continuing vicious cycle. Present thinking and culture survive thanks to a continual energetic degrading of our water, our food, our rivers, forests, natural environment and living conditions … that keep us thinking at a lower octave. This is another indication that we need to reach a

---

[325] Viktor Schauberger, Callum Coats, editor, *The Fertile Earth*, 147–48.

[326] Speaking to Ehrenfried Pfeiffer this is what Steiner expressed in relation to the weakness of the will in present time: 'That is a problem of nutrition. The form taken by nutrition today no longer gives people the strength to bring the spiritual to manifestation in physical life. They can no longer build the bridge from thinking to willing and action. The food plants no longer contain the forces they ought to be giving people.' (Ehrenfried Pfeiffer, '*Ein leben fur den Geist*', Perseus Verlag, Basel, 1999 quoted in Paul Emberson, *Machines and the Human Spirit: The Golden Age of the Fifth Kingdom*.)

tilting point in the scales in which cultural renewal will become a priority, over and above simple political reforms.

Callum Coats sums up the ideas of his teacher thus: 'In Viktor's view, the physical conditions of the human world and Nature are the direct, legitimate and inevitable outcome of humanity's spiritual concepts and ideological convictions.'[327] With this a spiritual-scientific view can only agree. Schauberger was such a holistic thinker that by necessity he had to converge toward the views of Goethe or Steiner. There was in his mind a direct connection between humanity's worldview and its spiritual content and the physical conditions in which human beings and Nature find themselves at present. This he could capture with images in which he saw larger implications than just seeming banalities. 'As long as the trout continues to stand motionless in the water because food flows unaided into its jaws, then favorable conditions will also exist for humanity and for the economy' is one such statement, which to surface thinking may appear an exaggeration.[328]

We have seen that this utmost respect of Nature which would allow the trout to survive can only be derived from a more holistic, but indeed very concrete understanding of the subtle forces that allow this fish to continue being part of its environment with practically no effort. When the trout has to become a beggar for its existence, as is mostly the case at present, the human being cannot lag far behind. No political programme can change this reality, if there is no willingness to see Nature in a more holistic way—we could say truly perceive Nature.

Recovering Balance, Countering Depletion

Schauberger was an adept of the 'both/and'. He naturally used his intuition through his unique ability to observe carefully. He was then able to understand through concepts, and apply these to his inventions. Synthesis was for him akin to natural, planetary motion; analysis to technical motion. He recognized that these two processes must constantly play out in equilibrium, though planetary motion needs to have the upper hand.

The Austrian forester felt that the balance between planetary motion (feminine) and technical motion (masculine) should be

---

[327] Callum Coats, *Living Energies*, 297.

[328] Viktor Schauberger, Callum Coats, editor, *The Water Wizard*, 158.

tilted toward the former in the ratio of 2/3 to 1/3; his son Walter felt it was according to the golden ratio Ø or 1.618 giving 61.8% to the feminine.[329] What this means in relation to planetary ecology and climate change is the reality of a constant degradation of Nature because not only what we think but what we put out in the world is of the nature of analysis/degradation rather than synthesis/upbuilding. The devastations imposed upon natural ecosystems because of the disregard of quality and type of energy and because of technology are presently of a scale comparable to what Nature itself produces. To come to a conclusion based on everything that has so far been explored, we could say that the balance between upbuilding forces and decaying forces has been altered in a way that a trained eye could see everywhere in the world. Such was the conclusion of Schauberger: 'the magnetism-generating Earth, which moves in cycloid-spiral-space-curves is now incapable of neutralizing the decomposive energies and their effects produced by millions and millions of human beings'.[330] Planetary depletion is therefore a much more inclusive term for the challenges humanity is facing than climate change. It's a terminology that addresses the effects not only on the climate, but on the very essence of life on Earth. This is how large the confrontation is.

Satisfying ourselves with the more prevalent views of climate change and its proposed solutions means, from a phenomenological/Goetheanistic perspective, paying lip service to a worldview which created the problem and cannot offer real solutions. If we can directly or indirectly observe in Nature what has been offered through the pages of this book—and it has been possible to this author—it will be clear that a view of climate change solely, or primarily, based on gas emissions and energy-production aspects comes short of addressing the global challenge.

The perspective that uses greenhouse emissions as the global culprit is deceptively self-serving; it leaves everything else off the hook. It allows change with minimal tinkering, while avoiding a hard look at the deeper drivers of planetary depletion. Ultimately it cannot succeed because it fails, or doesn't want, to see the larger picture. WWI and WWII have set the stage for a rampant and lawless economic globalization, and at present we reap the fruits of a complete reification of Nature, a ruthless exploitation that brings

---

[329] Alick Bartholomew, *Hidden Nature*, 53.

[330] Viktor Schauberger, Callum Coats, editor, *The Energy Evolution*, 79.

the Earth's regenerative capacities to a breaking point. A Goethean view of world ecology shows us the wreckage of such a worldview when it is left free to change the very nature of rivers, forests and farming, and much more. The water cycle has been impacted globally; so has the life energy of most natural ecosystems. The model that has taken hold of Nature worldwide continuously degrades the formation of finer energies, through which Nature is truly Nature; it depletes it of the etheric.

At present, not only the human being, but Nature itself is progressively cut off from the spirit, from its true source. Trees, crops and rivers are weaker and sicker worldwide because they are seen and treated as commodities. A view of Nature which sees only the quantitative has managed, to a great extent, to drive out quality from the environment and from the human being. In these circumstances Nature accomplishes the cleansing that comes with illnesses and catastrophes; it shows us the imbalances we have created.

In spiritual-scientific terms we risk bringing Nature to the level of sub-Nature, to estrange it from the spirit which is at the heart of Earth evolution, the Christ spirit. The elementals which wreak havoc upon Nature and climate have been won over by Ahriman. Abandoned by the human beings they may feel they have nowhere else to turn. Thus it is up to us to actively seek a true communion with Nature and the elemental beings, one which is eminently practical while it is also deeply spiritual. This is what we can discern in people like Goethe, Steiner, Hauschka, Schauberger and many others: a deeper connection with Self, with Nature and with the evolutionary impulse of the spirit of the Earth, the Christ spirit which ensouls it.

Only a true '*Kapieren* and *Kopieren*'—understanding Nature in order to act as it does—can present a lasting solution to global ecological challenges. Only a new culture can offer a remedy to an overall culture that has generated and perpetuated the problems, not just at the technical level, but primarily at the ideological one. There can be no solution at the level which has generated the problem, as Einstein would not tire to remind us.

We are facing a planetary choice, knowing that, yes, there is a danger of epic proportions, but on the other hand it is possible to thoughtfully perceive the reality of planet Earth, use resources in ways that render available clean and abundant energy, ennoble matter and improve the environment at the same time. We are truly finding ourselves in front of a great cultural divide. The way upward is a real departure from the past; it implies seeing humanity in a co-creator role with the divine. With good reason could Schauberger prophesy that: 'Human beings will thereby become true *creators*, who can so order the process of growth that the Earth will produce a superabundance of everything the increasing world-population needs in the way of food.'[331] This is in effect the other side of the equation, one that it is impossible to apprehend if we limit ourselves to addressing planetary depletion with political measures based on old thinking, even those predicated at a global scale. As Rudolf Hauschka would remind us: 'For our purpose theories are worse than useless. The world of ideas which comes to light in man must be brought to bear on his perceptions if he is to achieve real knowledge.'[332]

---

[331] Viktor Schauberger, Callum Coats, editor, *The Fertile Earth*, 150.
[332] Rudolf Hauschka, *The Nature of Substance*, viii.

Appendix

Turning Waste and CO_2 into Resources: The Blue Economy

What has been announced in Chapter 5 can now offer us a glimpse of how far we can go from the perspective of a Blue Economy toward implementing an economy of abundance that also addresses the climate/ecological crisis. We will look first at how energy and CO_2 can be addressed from a perspective of abundance, then turn our gaze to an uncommon experiment and example of Blue Economy consciously applied in an island environment, in spite of great limitations. In both instances we are really overcoming outdated thinking models, which are themselves the root cause of our present crisis. We are changing economic thinking from the very ground and turning waste into resource, addressing the climate challenge upstream rather than downstream.

From Waste to Abundance

We have already mentioned the project of Gaviotas in Colombia in relation to the regeneration of the tropical forest in Chapter 4. Gaviotas is located in Colombia's *llanos* on the western side of the Orinoco River. In the *llanos*—the savannah which has replaced the original forest—the pH is very low and therefore the water is not drinkable. Paolo Lugari and his collaborators discovered that the Caribbean pines (*Pinus caribaea*) inoculated with the mycorrhiza *Pisolithus tinctorius* were able to survive in this difficult environment, and alter the ecological conditions under the shade of their canopies. With the litter that they generate the trees improve the hydric regime, and lessen the temperature extremes of the soil horizon, attracting the original forest species no longer growing in the region. After new species grow, diversity returns and so does the rain. The water is once more potable. As we know, with a positive temperature gradient the water flows into the ground and replenishes the water table. At present water has returned to be abundant. It serves the need of the local 2,000 inhabitants and the excess is sold in the Bogota market. Gaviotas now collects resins from harvested trees which it processes with renewable energies and from which it offers nine different products. To cap it all, the waste goes into producing construction materials that close the loop. Over just above twenty years the improvement is such that land value has increased 3,000 times.[333] And another silent revolution has affected Colombia, as we will see shortly.

We have seen in Chapter 5 that only 0.2% of the coffee bush is enjoyed by the consumer, the rest is waste. Now, with what is left over the farmers can grow mushrooms and afford to raise livestock. This is possible because coffee produces a valuable hardwood and an ideal substrate for oyster and other mushrooms such as the highly quoted *reishi*, rich in medicinal properties. Mushrooms grow three times faster on caffeine-growth medium than on an oak substratum, their traditional one. Likewise, at the other end of the line, coffee grounds are an ideal growing medium for mushrooms, one that comes already sterilized.

Coffee hardly provides a livelihood to small farmers in years of high prices; during the rest of the time most growers and their families suffer. The Blue Economy has introduced the 'Pulp-to-Protein' idea with success in Colombia, Vietnam—the world's second exporter of coffee—Zambia and other countries. On the other hand coffee grounds are now exploited for mushroom production in Berlin and San Francisco. The two streams together can generate considerable means for improved livelihoods. The process is now well established and well documented by almost a decade of research and twenty scientific publications.

Under these new possibilities the Fair Trade label, even if better than organic alone, offers obvious limitations in terms of environmental stewardship. Adding the mushroom revenue to Fair Trade standards would strengthen sustainability and self-reliance on the part of the farmers.

In Colombia the state of El Huila already introduced the innovation in more than 100 companies by 2010, and this allows the state to replace illegal drugs with nourishing food.[334] Emilio Echeverri, former Vice-President of the Colombian Coffee Federation, who was an early advocate of this Pulp-to-Protein programme, became the governor of the coffee-growing state of Caldas.

---

[333] Gunter Pauli, *The Blue Economy: 10 Years, 100 Innovations, 100 Million Jobs.* Report to the Club of Rome, 17–18.

[334] Gunter Pauli, *The Blue Economy*, 94.

By 2010 there were already 10,000 people working with mushrooms in Colombia and dozens of villages exploiting the idea in Zimbabwe. India has followed suit, and training programmes have brought the technique to Tanzania, Congo, South Africa, Zambia, Mozambique, Cameroon.

The above are some of the most spectacular possibilities of using Blue Economy approaches to systemic change. We turn now to two promising technologies in terms of energy use and climate change: chemical free paper and cold process ceramics.

We have seen the tremendous generation of waste generated by the paper industry in Chapter 5. Nobel Laureate and Secretary of Energy for the Obama administration, Steven Chu, was interested in the termites and in addition their synergy with bacteria in processing wood. With a team he developed a technique that could reverse the need for wasteful pulping plants. The process has found healthy competition. The Chemistry Research Institute of Latvia has devised a pressure-based wood separation method that produces in succession hemicellulose, lignin and lipids and lastly separates cellulose as a clean residue. The two innovations together can create massive energy savings, eliminate waste and considerably reduce climate impact.[335]

Robert Ritchie of University of California, Berkeley has taken his inspiration from the abalone and *glycera* worm—commonly known as bloodworm. They both produce very hard ceramic composites at ambient temperatures, unlike common ceramics produced at high temperatures and presssure. The high-performance products that the process generates are stronger than the famous bulletproof Kevlar. With little operational overhead ceramics can be produced locally even on a small scale, allowing the flourishing of a multitude of businesses. High quality, energy-savings and positive climate impact all speak in favour of the technology.[336]

The above are just a few examples of what can be done when we imitate Nature and devise technological processes that tread lightly on the planet. But there is more. We are presently fortunate to be able to fathom the Blue Economy's potential when it is applied to a whole economic system. This has been done at the scale of an island.

---

[335] Gunter Pauli, *The Blue Economy*, 282–83.
[336] Gunter Pauli, *The Blue Economy*, 282.

Turning Limitations into Assets: the example of El Hierro

According to classical economic models all the island of El Hierro was deemed good for would have been a military training ground of one kind or another, leaving it deserted of most of its inhabitants. Instead, refusing to bend down to such a logic, it is now thriving.

El Hierro is an island of the Canary Archipelago that now fully illustrates the power of the Blue Economy. On the negative side it shows us that only in the periphery of the larger global economic system is a really sustainable economy allowed to flourish, and even then such an example is hardly touted for its positive impact by the forces that speak about fighting climate change. On the positive side we can see how much the determined will of the residents of the island, and their capacity to involve everyone in a deep listening process, can reverse the ravages of a globalized economy and offer an example of hope which does not require heavy mandates or control mechanisms, quite the contrary: transparency, inclusion and creativity. The confinement of the island becomes an asset in terms of showing the potentials of the Blue Economy in addressing climate change and an economy of scarcity.

El Hierro, the *'Isla Chiquita'* (little island) of Canary, covers 268 km², counts a stunning 500 volcano cones, and hosts 11,000 inhabitants. It has loudly proclaimed the goal of reaching energetic self-sufficiency through completely clean power within the next four to eight years. This is the result of a decision taken in 1997, when local resident Javier Morales contacted Gunter Pauli and the two agreed on a long-term collaboration, which is now starting to yield an abundant harvest.[337] The island initially adopted nineteen of the one hundred best innovations listed in the book, *The Blue Economy: 10 Years, 100 Innovations, 100 Million Jobs*. By 2006 already the island had achieved 82% of its overall objective policies, not just in terms of energy.[338]

The *Isla Chiquita* embarked on the ambitious goal of a regenerative economy resting on an 'economy of scope', trying to maximize synergies and interdependencies of enterprises building upon each other's resources and processes, where the waste of one becomes the resource of the other. Add to this the important caveat

---

[337] See http://www.zeri.org/ZERI/Home.html

[338] See https://www.hellocanaryislands.com/experiences/el-hierro-the-sustainable-island/

that companies are not allowed to export the profits outside of the island.[339]

In a systems-thinking approach to change, El Hierro has established a zero waste objective leaning on organic farming and sustainable fishing, renewable energy, sustainable public and private transportation, ecological tourism, entrepreneurship and the promotion of cooperative businesses.

For starters El Hierro's energy supply comes from the all too obvious wind energy. Added to these is an old crater converted into a water basin, exploiting the energy potential of height difference. The water stored in the crater is at the same time desalinated. In essence the water acts as a resource and as an energy battery. To avoid conflicts of interest water and energy management have been joined into one entity. The Gorona del Viento windmill farm, inaugurated in 2014, supplies 60% of the island's energy needs. It saved the island the equivalent of some 6000 tons of diesel in 2017.[340] To this will be shortly added wave power, exploiting the movement of ocean currents. To ensure that the proceeds of industry generation remain in the island the islanders own 60% of the enterprise in partnership with Endesa—the large Spanish energy company—the ITC Group and the Canary Islands Government. The proceeds from energy generation thus remain on the island.

El Hierro is improving the system of public transportation and especially the service to its elders, adding electric minibuses to its services. At the private level it is encouraging a car-share pilot programme for public employees, shared use of taxis, the use of electric cars or of those powered by biodiesel from vegetable oils.

As the island creates income from farming, conversion to organics has been quite effective, through the use of Terra Preta—a process that adds a mixture of charcoal, bones, broken pottery, compost, and manure to the organic material—and other composting methods. The island produces wine, bananas, papayas, pineapples, feed for animals, goat cheese and yogurt, and has vowed to recycle all that is left from consumption. El Hierro counts upwards of 4000 hectares in production, much of

---

[339] For a study of the regenerative economy of the island see the article by Desirée Driesenaar at https://medium.com/age-of-awareness/heroes-of-el-hierro-part-ii-f5a25937a21b

[340] See https://www.hellocanaryislands.com/experiences/el-hierro-the-sustainable-island/

it already organic. It also inaugurated the PapaGaya Permaculture School encouraging a number of projects on the island. With the advantage of a short food supply chain farmers get a good price for their produce.

Part of the food supply obviously comes from fishing and improving this industry has been a focused objective. To this effect the locals have safeguarded a section of coastal area as a sanctuary for fishstock regeneration and have moved operations to fishing with lines rather than nets. The tuna catch is available to the islanders, then to the rest of the Canary Islands, and what is left over goes to Spain. To cap it all, tourism, Canary's traditional industry, offers a complement to all the other outlets through a specialized niche for the mindful explorer.

Bibliography

Addey, John, *Harmonics in Astrology: An Introductory Textbook to the New Understanding of an Old Science*

Alexandersson, Olof, *Living Water: Viktor Schauberger and the Secrets of Natural Energy*

Bartholomew, Alick, *Hidden Nature: The Startling Insights of Viktor Schauberger*

Blattmann, Georg, *The Sun: The Ancient Mysteries and a New Physics*

Coats, Callum, *Living Energies: Viktor Schauberger's Brilliant Work with Natural Energy Explained*

Cobbald, Jane, *Viktor Schauberger: A Life of Learning from Nature*

Edwards, Lawrence, *The Vortex of Life: Nature's Patterns in Space and Time*

Emberson, Paul, *Machines and the Human Spirit: The Golden Age of the Fifth Kingdom*

Emoto, Masaru, *Messages from Water*

Hauschka, Rudolf, *The Nature of Substance: Spirit and Matter*

Koonin, Steven E., *Unsettled: What Climate Science Tells Us, What It Doesn't, and Why It Matters*

Kronberger, Hans and Lattacher, Siegbert, *On the Track of Water's Secrets: from Viktor Schauberger to Johanner Grander*

Landscheidt, Theodor,

- *Sun-Earth-Man: a Mesh of Cosmic Oscillations: How planets regulate solar eruptions, geomagnetic storms, conditions of life, and economic cycles*
- *Solar Activity: A Dominant Factor in Climate Dynamics*
- *The Golden Section: A Cosmic Principle*

Marti, Ernst, *The Four Ethers: Contributions to Rudolf Steiner's Science of the Ethers*

McCormac, Billy M., editor, *Weather and Climate Responses to Solar Variations*

Mörner, Nils-Axel, *The Approaching New Grand Solar Minimum and Little Ice Age Climate Conditions*

Moseley, Michael E., *The Incas and Their Ancestors: The Archaeology of Peru*

Pauli, Gunter, *The Blue Economy: 10 Years, 100 Innovations, 100 Million Jobs; Report to the Club of Rome*

Quinn, W., *Registros del Fenómeno El Niño en el Perú*

Scafetta, Nicola, *Testing an astronomically based decadal-scale empirical harmonic climate model versus the IPCC (2007) general circulation climate models*

Scafetta, Nicola and Bianchini, Antonio, *Overview of the Spectral Coherence between Planetary Resonances and Solar and Climate Oscillations*

Schauberger, Viktor and Coats, Callum, editor,

- *The Water Wizard: The Extraordinary Properties of Natural Water*
- *Nature as Teacher: New Principles in the Working of Nature*
- *The Fertile Earth: Nature's Energies in Agriculture, Soil Fertilisation and Forestry*
- *The Energy Evolution: Harnessing Free Energy from Nature*

Schwuchow, Jochen, Wilkes, John, Trousdell, Iain, *Energizing Water: Flowform Technology and the Power of Nature*

Shady, Ruth and Leyva, Carlos, editors., *La ciudad sagrada de Caral-Supe: las orígenes de la civilización andina y la formación del Estado prístino en el antiguo Perú*

Silva, Freddy, *Secrets in the Field: The Science and Mysticism of Crop Circles*

Steiner, Rudolf,

- *The Foundations of Esotericism*
- *The Festivals and Their Meaning*
- *Three Streams in the Evolution of Mankind*
- *Man as a Symphony of the Creative Word*
- *Anthroposophical Leading Thoughts: Anthroposophy as a Path of Knowledge; The Michael Mystery*

Taylor, Peter, *Chill: A Reassessment of Global Warming Theory*

Tompkins, Peter and Bird, Christopher, *Secrets of the Soil: A Fascinating Account of Recent Breakthroughs—Scientific and Spiritual—That Can Save Your Garden or Farm*

Vijaya, Gopi Krishna,

- *Replacing the Foundations of Astronomy*
- *Celestial dynamics and rotational forces in circular and elliptical motions*
- *Unlocking Eclipses and Planetary Distances*

Wachsmuth, Guenther, *The Etheric Formative Forces in Cosmos, Earth and Man: A Path of Investigation into the Living World*

Internet Sources

https://www.worldatlas.com/articles/the-deadliest-dam-failures-in-history.html

https://en.wikipedia.org/wiki/Dam_failure#List_of_major_dam_failures

https://www.geekwire.com/2022/heres-a-look-at-the-worlds-biggest-carbon-capture-site-which-just-signed-a-10–year-deal-with-microsoft/

https://www.oxy.com/news/news-releases/occidental-and-blackrock-form-joint-venture-to-develop-stratos-the-worlds-largest-direct-air-capture-plant/

https://www.reuters.com/business/energy/occidental-ceo-sees-potential-license-1000–carbon-capture-plants-2023–11–08/

https://www.weforum.org/agenda/2023/08/how-to-get-direct-air-capture-under-150–per-ton-to-meet-net-zero-goals/

https://www.iea.org/reports/direct-air-capture-2022

https://www.ipcc.ch/report/ar6/wg2/downloads/report/IPCC_AR6_WGII_SummaryForPolicymakers.pdf

https://www.ipcc.ch/report/ar6/wg2/downloads/report/IPCC_AR6_WGII_TechnicalSummary.pdf

'Pattern Recognition in Physics', *'General Conclusions regarding the Planetary-Solar-Terrestrial Interaction'*, journal of 2013, available at http://www.pattern-recogn-phys.net/1/203/2013/prp-1–203–2013.pdf

https://www.oxy.com/news/news-releases/occidental-and-blackrock-form-joint-venture-to-develop-stratos-the-worlds-largest-direct-air-capture-plant/

https://thehill.com/changing-america/sustainability/climate-change/3669378–the-worlds-largest-carbon-removal-project-will-break-ground-in-wyoming/.

https://www.geekwire.com/2022/heres-a-look-at-the-worlds-biggest-carbon-capture-site-which-just-signed-a-10–year-deal-with-microsoft/

https://www.weforum.org/agenda/2023/08/how-to-get-direct-air-capture-under-150–per-ton-to-meet-net-zero-goals/

https://www.iea.org/reports/direct-air-capture-2022

http://www.zeri.org/ZERI/Home.html

https://www.hellocanaryislands.com/experiences/el-hierro-the-sustainable-island/

https://medium.com/age-of-awareness/heroes-of-el-hierro-part-ii-f5a25937a21b